RAILWAYS RESTORED

1991/2 EDITION

EDITED BY ALAN C. BUTCHER

LONDON

IAN ALLAN LTD

Chairman's Foreword

Last year witnessed some mixed fortunes in the railways preservation world. It was notable for the celebration of 40 and 30 years preservation of narrow and standard gauge respectively in the guise of the Tallylyn, Bluebell and Middleton Railways.

Britain, and ARPS in particular, is still regarded as the world leader in the world of railway preservation. The following pages will explain why. However, we may not rest on our laurels as Continental steam railways and centres are burgeoning all over Europe, who are showing an increasing interest in what we're doing, particularly those from Central Europe, previously behind the Iron Curtain, who attended the ARPS meetings in Utrecht, Sittingbourne and on the 'Bluebell'. During 1991 the ARPS will again be holding a meeting on the Continent, this time at Bruckhausen-Vilsen, between Bremen and Hanover, which will be followed by visits to other steam operations including at least one in eastern Germany.

1990 saw the closure of the Dinting Railway Centre, which was born again in revitalised form as the Bahamas Locomotive Society at Ingrow on the Keighley & Worth Valley Railway. Indeed, their Good Samaritans' act was largely instrumental in the ARPS Annual Award being granted to the Worth Valley jointly with the Vintage Carriage Trust. Ingrow is fast becoming a showpiece of railway preservation and now justifies a special visit in its own right.

Passenger and visitor figures also seem to have varied from one railway to another, without any clear indication or factor as to the reason. The West Somerset Railway enjoyed another record year while the doubts over the viability of the Buckfast Line persuaded its parent company, the Dart Valley Railway PLC, to hand over the operation of the line between Buckfastleigh and Totnes to the Dart Valley Railway Supporters' Association, thus resulting in more extensive use of unpaid volunteers.

Most railways do indeed encourage the participation of their supporters. Do not be deluded by the seemingly effortless and efficient running of the railways you visit into believing that your help would not be appreciated. If you are not already a member of a railway preservation society, join one that catches your fancy and discover what a rewarding interest railways can be. If you are unsure which to join, become an individual member of the ARPS itself. You will be very welcome.

David Morgan
Chairman ARPS

Contents

First published 1991

ISBN 0 7110 2000 0

Published by Ian Allan Ltd, Shepperton, Surrey and printed by Ian Allan Printing Ltd at their works at Coombelands in Runnymede, England

Editor's Notes

On the following pages will be found a guide to the major preserved railways, railway museums and preservation centres in the British Isles. Information for visitors has been set out in tabular form for easy reference, together with a locomotive stocklist for most centres.

Many preservation centres and operating lines provide facilities for other groups and organisations to restore locomotives and equipment on their premises. It has not been possible to include full details of these groups, but organisations which own locomotives are shown under the centres at which they operate. In addition a full list of member societies of the ARPS is given elsewhere. In the case of most operating lines their length is given but there is no guarantee that services are operated over the entire length.

Within the heading to each entry a heading block has been incorporated for easy reference as to what each site offers in the way of passenger service to visitors. These are as follows:

Railways providing a passenger service between two or more stations with public access; eg Mid-Hants Railway.

SC A railway or preservation site offering a passenger service on a short length of line, on a regular basis, with public access at only one point; eg Lavender Line.

M A museum or site that does *not* offer a passenger service on a regular basis, if at all; eg Science Museum, London. Some sites may however offer rides on miniature railways.

Under the heading **References** will be found the details of articles which have appeared in the Ian Allan magazine *Railway World* during the past two years.

Details given under **Access by public transport** should be checked beforehand to ensure services shown are operating.

Alan C. Butcher

Frontcover:
The beautifully restored Wigan Coal & Iron Railway locomotive *Lindsay* in steam at Carnforth on 3 November 1990. *Mike Taylor*

Introduction

The Preserved Line Phenomenon

The ARPS have now been at it for 30 years . . . and the Talyllyn for 40! It is said that only fishing and football rate higher than railway enthusiasm as Britain's most popular male leisure sport.

I am fortunate in having been involved in standard gauge preservation from the very outset, and have watched the movement grow from the humble beginnings of the Middleton and Bluebell railways into the major network we enjoy today.

Railway preservation is a classless hobby. People from every conceivable walk of life rub shoulders on track gangs and in locomotive workshops. It is also a hobby that can be enjoyed by people of almost any age. On one railway's p-way gang, a teenager and a septuagenarian regularly work together. And it is a rewarding hobby. Few others give their participents so much to show for their efforts at the end of it all.

However, it is a hobby that is beset with violent differences of opinion. One would think that with so many clergymen within its ranks, everyone would agree with each other and live in peace, but they don't. (In fact neither do the clergymen.)

Some preservationists are the most bigoted, opinionated men one could ever come across, and get very aggressive when 'their own' particular locomotive, line or livery is called into question, or 'their' railway does not do what they think it should.

It certainly has its origins within the 'territorial' influences of the Big Four (now defunct) and the BR regions (now replaced by business sectors), but whatever the reasons it has caused several 'mutinies'. There were three big ones in earlier years, the M&GN Society, the Kent & East Sussex and the Mid-Hants. On occasions the ARPS has had to go in and bash their heads together but they nearly always come out stronger.

Ironically, it is these emotions which have proved the dynamism and energy behind our achievements and they are one of the main reasons why the movement has achieved so much in the past 30 years.

For, if apathy reigned, we would never have got off the ground.

Preservation is both a disciplined and undisciplined service at one and the same time. That's one of its great attractions. We don't have a creed but if we did it would probably concern 'loyalty to one's railway or society'. For this unshakable passion by a volunteer for his, or her, favourite line or locomotive has proved to be one of the enduring strengths when the going is tough . . . and 1991 looks tough.

It is a bit like the Navy. A ship might have an absolutely despised captain but the crew's loyalty is not to him, but the ship.

Our achievements are now a legend, but to reach this stage, we have had to put up with a lot of doubt and cynicism. A good case in point is Barry. How many times did we hear in the early days that the restoration of this or that locomotive was impossible and the cry of the smug Jonahs 'enough is enough'.

Well, the word impossible has become rather common because 'impossible' to most people outside our world is merely considered 'difficult' by us.

And now Barry has won a prestigious award from British Coal's steam heritage.

I think the NRM's greatly loved chief mechanical engineer summed it up just before he died in 1988 when John Bellwood wrote to Francis Blake of Barry Rescue 'whilst the engineer in me used to say enough is enough and that the best use of Dai Woodham's loss-making investment was as spares for other preserved locos, there was an inner voice saying "where there's a hulk, there's a hope for another working loco". After 71000 anything is possible.'

For someone like John to come round to that way of thinking showed how far we had come.

Strictly commercially Barry was a load of codswallop because nothing that came out of there could be considered financially viable. But railway preservation doesn't obey the normal rules of commerce — volunteer determination overcomes time limits and labour costs (and inflation).

Preserved railways must have a profit motive, naturally, but it must not be for *personal* profit. That's not to say that people shouldn't receive a salary but the whole secret of our phenomenal and sustained success is that very few people are in it for personal gain. If they were we would be in trouble.

A selfless attitude is one of two things at the heart of railway preservation. The other is fun. The minute it ceases to be fun it will be all over.

We are well prepared for stormy weather in 1991.

Peter Manisty
Vice President ARPS

The Ian Allan Railway Heritage awards

This year is the 12th of the Railway Heritage Awards. Originally intended to recognise and encourage good work in the repair and restoration of the preserved railways' stations, the scheme quickly expanded to include all kinds of railway structures in any ownership, covering meritorious restoration work by British Rail and, in the case of redundant railway buildings, those which have passed into other ownerships such as commercial firms, local authorities and private individuals. The main criterion is that the work must be sympathetic towards the structure's original purpose as a railway building, and reflect it as such.

Ian Allan Ltd are the principal sponsors, supported by the British Rail Community Unit, The Railway Heritage Trust and British Coal. The scheme is divided into two sectors, one of the voluntarily-operated railways and the other for publicly or commercially owned structures. Awards have been given for a very wide variety of schemes. The majority have been stations, but there have also been restored pumping engine houses, bridges, signalboxes and other features of the railways' infrastructure. In the public and commercial sector, particular emphasis is placed on encouraging the sympathetic and imaginative re-use of old railway buildings.

In alternate years an award is given in the voluntary sector for the best restored signalling installation on a preserved railway; 1991 is a 'signalling year'. From time to time awards are also given for other specific features in this sector, such as the best restored booking office and, in 1990, the best restored footbridge.

The awards are presented in London each March, at a prestigious ceremony attended by people prominent in the world of conservation and heritage affairs. The presentations are made by well known public figures, and have included Prince Michael of Kent, the Chairman of English Heritage Lord Montagu of Beaulieu and David Weston the well known artist.

The awards themselves are as follows:

Voluntary Sector

The British Rail Award for the most meritorious entry — an hand-inscribed and gilded slate plaque presented by British Rail Community Unit, with a cheque for £1,500 presented jointly by British Rail Community Unit and Ian Allan Ltd.

First Class Awards for particularly outstanding entries — not more than three die-cast metal plaques presented by ARPS, together with cheques for £750 presented by Ian Allan Ltd.

The Best Restored Signalling Award — a certificate of commendation and a cheque for £250 presented by British Coal, and awarded in alternate years — including 1991.

The Railway World *Commendation* — a plaque and a grant of £1,000 presented by *Railway World* magazine for an incomplete project showing particularly outstanding promise. This award is intended to give tangible encouragement to projects which still have some way to go towards completion.

Certificate of Commendation awarded from time to time for a special feature.

Public and Commercial Sector

The Ian Allan Award for the most meritorious entry — an inscribed brass plaque presented by Ian Allan Ltd.

First Class Awards — not more than three die cast metal plaques presented by ARPS for particularly outstanding entries.

Certificates of Commendation for entries of high merit or showing particular degree of promise.

Either Sector

The Railway Heritage Trust Award — a metal plaque for the best restored listed structure, or structure of special historical interest, in which there has been involvement by the Trust.

Special Commendations in the form of certificates which may be awarded at the discretion of the adjudicators in special circumstances as they think fit.

The awards scheme is organised by ARPS, and judges are drawn from a country-wide panel having special expertise in various appropriate fields. Entries gaining a high initial mark are seen a second time independently by a second judge, and the final choice is made by a panel of seven adjudicators in November, in readiness for the presentations in the following March. Entries are placed in one of four classes, and marks are given for exterior, interior and environmental treatment, and for users' and visitors' facilities, broken down into a range of different features as appropriate to the type of entry.

Entry forms may be obtained from the Awards Secretary, Arthur Harding, 6 Ullswater Grove, Alresford, Hants SO24 9NP, telephone 0962 733327.

Gordon Biddle

Association of Independent Railways

The Association of Minor Railways (AMR) was formed at the time when railways were fighting for a 'Square Deal' with road competition, and those not parties to the Railway Companies Association formed the AMR in 1938. Its scope and objects were defined then and have remained primarily as a means of combined direct representation to Government departments, as well as discussing matters of mutual interest. It was recognised by the Ministry of Transport in 1939.

Meetings became very infrequent during the war and many of its members were nationalised, and meetings virtually ceased.

In 1944 its membership was limited to those railways operating under the Acts or Light Railway Orders.

During the 1960s the managers of the longer established railways felt that it would be useful if they could meet more formally from time to time to discuss matters of common interest and the AMR was revived in 1970.

Since then it has grown in status with considerably larger attendances at twice yearly general meetings when principal directors and officers take the opportunity of discussing matters of mutual interest and on deciding matters of general policy.

The Association's day to day management is in the hands of a committee of six elected representatives, a Secretary and Treasurer. In April 1988 it became a company limited by guarantee and the name changed to the Association of Independent Railways Ltd.

The objective of representing the industry to Government and other bodies remains unchanged as does the intention to encourage mutual co-operation — not at the expense of fair and healthy competition — between its member companies.

The ARPS Annual Award

This, the premier award made by ARPS, is made to a group or organisation making an outstanding contribution to railway preservation during the year of the award.

The Award takes the form of a Royal Train Headboard from the London, Brighton & South Coast Railway, which is on loan to ARPS from the National Railway Museum. The award is held for one year and the winning group also receives a commemorative plaque. The Award is announced and presented at the Association's Annual General Meeting which is held on the last weekend of January each year.

The award for 1990 is made jointly to the Keighley & Worth Valley Railway and the Vintage Carriages Trust 'for successful and practical redevelopment of Ingrow station to include homes for the Vintage Railway Carriage Museum and the Bahamas Locomotive Society'.

The human side of the award should not be forgotten in that the K&WVRPS made space available on an already overcrowded line to house both societies and their rolling stock.

Previous awards have been made to:

1975—the Great Western Society Ltd
1976—the Princess Elizabeth Locomotive Society
1977—the Peterborough Railway Society Ltd (Nene Valley Railway)
1978—the Festiniog Railway Society
1979—the Severn Valley Railway Co Ltd
1980—the Scottish Railway Preservation Society Co Ltd
1981—the Great Western Society Ltd
1982—the Festiniog Railway Society Ltd
1983—the Railway Preservation Society of Ireland
1984—the Severn Valley Railway Co Ltd
1985—jointly to the A4 Locomotive Society Ltd Merchant Navy Locomotive Preservation Society Ltd and the Friends of the National Railway Museum
1986—the 71000 Duke of Gloucester Locomotive Trust Ltd
1987—the East Lancashire Railway Society
1988—staff and Friends of the National Railway Museum
1989—the Kent & East Sussex Railway

Abbreviations

AEC	Associated Equipment Co
AEG	Allgemeine Eletricitaets Gesellschaft
A/Porter	Aveling & Porter Ltd
A/Whitworth	Armstrong Whitworth
B/Drewry	Baguley/Drewry
B/Peacock	Beyer Peacock & Co
B/Hawthorn	Black, Hawthorn & Co
BTH	British Thomson Houston
D/Metcalfe	Davies & Metcalfe
E/Electric	English Electric Ltd
F/Jennings	Fletcher Jennings & Co
F/Walker	Fox Walker
G/England	George England & Co
H/Clarke	Hudswell Clarke & Co Ltd
H/Leslie	Hawthorn Leslie & Co
H/Hunslet	Hudson Hunslet
K/Stuart	Kerr Stuart & Co Ltd
M/Rail	Motor Rail Ltd
M/Vick	Metrovick (Metropolitan-Vickers)
M/Wardle	Manning Wardle & Co Ltd
N/British	North British Locomotive Co Ltd
N/Wilson	Nasmyth Wilson & Co Ltd
O&K	Orenstein & Koppel
RSH	Robert Stephenson & Hawthorn Ltd
R/Hornsby	Ruston Hornsby
R/Proctor	Ruston Proctor
YEC	Yorkshire Engine Co
DB	German Federal Railways
DSB	Danish State Railways
JZ	Jugoslovenske Zeleznice
MoS	Ministry of Supply
NSB	Norwegian State Railways
RR	Rhodesian Railways
SJ	Swedish Railways
SAR	South African Railways
SNCF	French National Railways
USA TC	United States Army Transportation Corps
WD	War Department
BE	Battery electric
DE	Diesel-electric
DH	Diesel-hydraulic
DM	Diesel-mechanical
E	Overhead electric
F	Fireless
G	Geared
PM	Petrol-mechanical
ParM	Paraffin-mechanical
PT	Pannier tank
R	Railcar
ST	Saddle tank
T	Side tank
VB	Vertical boiler
WT	Well tank
4w	4-wheel

South East

Amberley Chalk Pits Museum — SC

Narrow Gauge and Industrial Railway Collection (incorporating the Brockham Museum of Narrow Gauge Railways)

The NG&IR Collection is part of an open air industrial museum set in 36 acres of the former Pepper & Co chalk pits. A 2ft gauge line has been constructed and this is used for carrying passengers in genuine workmen's vehicles

Location: Houghton Bridge, Amberley, West Sussex (3 miles north of Arundel). On B2139. Adjacent to Amberley BR(SR) station

OS reference: TQ 031122

Operating society/organisation: Southern Industrial History Centre Trust, Amberley Chalk Pits Museum, Houghton Bridge, Amberley, Arundel, West Sussex BN18 9LT

Telephone: Bury (0798) 831370 (Museum office)

Car park: Adjacent to Amberley Station

On site facilities: Shop, light refreshments, audio-visual show

Public opening: Wednesday to Sunday (inclusive) each week, and Bank Holiday Mondays, (open all week in school summer holidays) 10.00-last entry 17.00, 20 March-27 October

Special events: Please see press for details

Special Notes: The NG&IR Collection is part of an open air industrial museum set in 36 acres of the former Pepper & Co chalk pits. Displays include working potter, blacksmith, boatbuilder and printer, stationary engines, historic radio collection and newly completed vintage Southdown garage and buses. A 2ft 0in gauge industrial railway system is demonstrated when possible, and a 3ft 2¼in gauge line is under construction. In addition a 2ft 0in gauge 'main line' has been constructed and this is used for carrying passengers in genuine

Locomotives
(2ft or 60cm unless otherwise indicated)

Name	*No*	*Builder*	*Type*	*Built*
Polar Bear	—	Bagnall (1781)	2-4-0T	1905
Peter	—	Bagnall (2067)	0-4-0ST	1918
Lion	—	Baldwin (44656)	4-6-0T	1917
Townsend Hook	4	F/Jennings (172L)	0-4-0T (3ft 2¼in gauge)	1880
Scaldwell	—	Peckett (1316)	0-6-0ST (3ft 0in gauge)	1913
—	23†	Spence	0-4-0T (1ft 10in gauge)	1921
—	—	Decauville (1126)	0-4-0T	1950
—	740	O&K (2343)	0-6-0T	1907
Monty	(6)	O&K (7269)	4wDM (3ft 2¼in gauge)	1936
The Major	(7)	O&K (7741)	4wDM	1937
—	2	Ransomes & Rapier (80)	4wDM	1937
—	—	Hudson-Hunslet (3097)	4wDM	1944
—	2	R/Hornsby (166024)	4wDM	1933
—	(3041)	M/Rail (Simplex) (1320)	4wDM (ex-PM)	1918
—	3101	M/Rail (Simplex) (1381) Armoured	4wPM	1918
Peldon	—	John Fowler (21295)	4wDM	1936
Redland	—	O&K (6193)	4wDM	1937
—	—	Lister (35421)	4wPM	1949
—	(LR 2593)	M/Rail (Simplex) (872)	4wPM	1918
—	27	M/Rail (Simplex) (5863)	4wDM	1934
—	—	M/Rail (Simplex) (10161)	4wDM (2ft 11in gauge)	1949
Ibstock	—	M/Rail (Simplex) (11001)	4wDM	1951
CCSW	—	Hibberd (1980)	4wDM	1936
Thakeham Tiles	No 3	Hudson-Hunslet (2208)	4wDM	1941
Thakeham Tiles	No 4	Hudson-Hunslet (3653)	4wDM	1948
—	—	H/Clarke (DM686)	0-4-0DM	1948
Star Construction	—	Hudson-Hunslet	4wDm	c1941
—	18	R/Hornsby (187081)	4wDM	1937
—	—	Hudson-Hunslet (2536)	4wDM	1941
—	—	Lister (33937)	4wDM	1949
—	—	R/Hornsby (172892)	4wDm	1934
—	WD 904	Wickham (3403)	2w+2PMR	1943

workmen's vehicles. The 500yd line, was officially opened by HRH Prince Michael of Kent on 5 June 1984. The railway is operated every day the museum is open (subject to mechanical availability), with steam locomotive haulage on certain days — for details contact the museum office. Wheelchairs can normally be accommodated on the train. A Narrow Gauge Industrial Railway Introductory Exhibition sets the scene for these and other set-piece display areas

Membership details: Friends of the Amberley Chalk Pits Museum, c/o above address

—	2	Wingrove & Rogers (5031)	4wBE	1953
—	—	Wingrove & Rogers (5034)	4wBE	1953
—	—	Wingrove & Rogers (4998)	4wBE	1953
—	—	Wingrove & Rogers (T8033)	0-4-0BE	1979

Stock

2 Penrhyn Quarry Railway 4-wheel coaches (2ft gauge, ex-1ft 10¾in gauge)
RAF Fauld bogie coach (1940) (2ft gauge)
Rye & Camber Tramway bogie (incomplete) (1895) (3ft gauge)
2 Groudle Glen Railway 4-wheel coaches (1896 and 1905) (2ft gauge)
60 other varied pieces of rolling stock of 12 different gauges ranging from 1ft 6in to 3ft 2¼in plus numerous miscellaneous exhibits including track, signals etc

†Includes hoist and 'haulage truck' for conversion to 5ft 3in gauge

Bluebell Railway

This famous steam railway was the first standard gauge passenger line to be taken over by enthusiasts. It derives its name from bluebells which proliferate in the woodlands adjoining the line. A strong Victorian atmosphere pervades this branch line which has a large collection of Southern and pre-Grouping locomotives and coaches.

Headquarters: Bluebell Railway Preservation Society, Sheffield Park station, Uckfield, East Sussex TN22 3QL

Telephone: Newick (082 572) 2370 for travel information (24hr talking timetable); Newick 3777 for bookings etc during office hours

Main station: Sheffield Park

Other public stations: Horsted Keynes

Car parks: Sheffield Park, Horsted Keynes

OS reference: Sheffield Park TQ 403238, Horsted Keynes TQ 372293

Access by public transport: Haywards Heath BR and bus to Horsted Keynes village weekdays, or direct to Sheffield Park on summer and winter Sundays. Additional bus services on certain days, timetable brochure available by sending sae

Refreshment facilities: Sheffield Park, Horsted Keynes (licensed bars at both stations). The 'Regency Belle' wine and dine train runs on certain dates and is also available for private charter

Souvenir shops: Sheffield Park, Horsted Keynes

Locomotives

Name	*No*	*Origin*	*Class*	*Type*	*Built*
—	27	SECR	P	0-6-0T	1910
Stepney	55	LBSCR	A1X	0-6-0T	1875
Fenchurch	72	LBSCR	A1X	0-6-0T	1872
Pioneer II	1178	SECR	P	0-6-0T	1910
Bluebell	323	SECR	P	0-6-0T	1910
—	263	SECR	H	0-4-4T	1905
Birch Grove	473	LBSCR	E4	0-6-2T	1898
—	488	LSWR	0415	4-4-2T	1885
—	592	SECR	C	0-6-0	1901
—	58850	NLR	2F	0-6-0T	1880
Earl of Berkeley	3217	GWR	9000	4-4-0	1938
—	30064	SR	USA	0-6-0T	1943
—	541	SR	Q	0-6-0	1939
—	830*	SR	S15	4-6-0	1927
—	847	SR	S15	4-6-0	1937
Stowe	928	SR	V	4-4-0	1934
—	1618	SR	U	2-6-0	1928
—	31638	SR	U	2-6-0	1931
—	96	LSWR	B4	0-4-0T	1905
—	C1	SR	Q1	0-6-0	1942
Sir Archibald Sinclair	34059	SR	BB	4-6-2	1947
Port Line	35027	SR	MN	4-6-2	1948
Camelot	73082	BR	5MT	4-6-0	1955
—	75027	BR	4MT	4-6-0	1954
—	78059†	BR	2MT	2-6-0	1956
—	80064	BR	4MT	2-6-4T	1953
—	80100	BR	4MT	2-6-4T	1954
—	92240	BR	9F	2-10-0	1959
Blackmore Vale	21C123	SR	WC	4-6-2	1946

Industrial locomotives

Name	*No*	*Builder*	*Type*	*Built*
Blue Circle	—	A/Porter (9449)	2-2-0TG	1926
Baxter	—	F/Jennings (158)	0-4-0T	1877
Stamford	—	Avonside (1972)	0-6-0ST	1927
Sharpthorn	—	M/Wardle (641)	0-6-0ST	1877

*Purchased without tender
†Purchased without tender, for conversion to tank engine

Museum: Sheffield Park
Depots: Sheffield Park (locomotives), Horsted Keynes (stock)
Length of line: 5 miles — plus first section of extension of route to East Grinstead, with shuttle service at weekends and some weekdays as notified
Passenger trains: Sheffield Park-Horsted Keynes
Period of public operation: Sundays in January, February and December, also Boxing and New Year's Day. Saturdays and Sundays in March, April and November. Wednesdays, Saturdays and Sundays in May and October. Daily June-September inclusive. Daily for October schools half term week. Museum and locomotive sheds at Sheffield Park open daily except Christmas Day
Special events: 21 April — Fire engine day; 12 May — Parade day; Vintage Sunday, in September together with various other events detailed in timetable brochure (SAE for details)
Facilities for disabled: All station facilities are on the level and ramps available for placing wheelchair visitors into trains. Special toilet in Buffet at Sheffield Park
Membership details: Membership Secretary c/o above address
Feature article reference: RW September 1990, p558

Stock

Substantial collection of pre-Nationalisation coaches including SECR, LSWR, Bulleid, Maunsell and Chesham vehicles. Also freight stock and engineers' vehicles plus 45ton steam crane

Owners

592 (the Wainwright C Class Preservation Society)
541, 830, 847 and 1618 (the Maunsell Locomotive Society Ltd)
96 and 21C123 (the Bulleid Society Ltd)
263 (the H Class Trust)
73082 (the Camelot Locomotive Society)
C1 (on loan from the National Railway Museum)
928 (on loan from Montagu Venturers Ltd)
80064 (80064 Group)
35027, 1178 (the *Port Line* Project)

Kent & East Sussex Steam Railway

The epitome of the early Edwardian light railways that were developed on shoestring budgets to open up less populated areas of the countryside. The line is now being extended towards Bodiam Castle and possibly to Robertsbridge providing an interchange with Network SouthEast. The picturesque town of Tenterden remains the headquarters of the railway.
Headquarters: Tenterden Railway Co Ltd, Tenterden Town station, Tenterden, Kent TN30 6HE
Telephone: Tenterden (058 06) 2943 (24 hour talking timetable); Tenterden
Main station: Tenterden Town
Other public stations: Rolvenden, Wittersham Road, Northiam
Car parks: Tenterden, Wittersham Road, Northiam
OS reference:
Tenterden TQ 882336,
Rolvenden TQ 865328,
Northiam TQ 834266
Access by public transport: Maidstone & District bus service No 400 from Ashford (Kent) BR station
Refreshment facilities: Tenterden Town. Also on train facilities. Lunch and afternoon teas on trains by prior arrangement
Souvenir shop: Tenterden Town station
Museum: Station Road, Tenterden
Depot: Rolvenden
Length of line: 7 miles

Locomotives

Name	*No*	*Origin*	*Class*	*Type*	*Built*
—	32670	LBSCR	A1X	0-6-0T	1872
Sutton	10 (32650)	LBSCR	A1X	0-6-0T	1876
Knowle	32678	LBSCR	A1X	0-6-0T	1880
—	1556	SECR	P	0-6-0T	1909
Wainwright	21 (30070)	SR	USA	0-6-0T	1943
Maunsell	30065	SR	USA	0-6-0T	1943
—	20	GWR	AEC	diesel railcar	1940
—	19	NSB	21c	2-6-0	1919
—	44 (D2023)	BR	03	0-6-0DM	1958
—	45 (D2024)	BR	03	0-6-0DM	1958
—	46 (D2205)	BR	04	0-6-0DM	1958
—	D9504	BR	14	0-6-0DH	1964
—	49 (D9525)	BR	14	0-6-0DH	1965

Industrial locomotives

Name	*No*	*Builder*	*Type*	*Built*
Marcia	12	Peckett (1631)	0-4-0T	1923
Charwelton	14	M/Wardle (1955)	0-6-0ST	1917
Holman F. Stephens	23	Hunslet (3791)	0-6-0ST	1952
William H. Austen	24	Hunslet (3800)	0-6-0ST	1953
Northiam	25	Hunslet (3797)	0-6-0ST	1953
Linda	26	Hunslet (3781)	0-6-0ST	1952
Rolvenden	27	RSH (7086)	0-6-0ST	1943
—	40	BTH	Bo-Bo	1932
—	42	Hunslet (4208)	0-6-0DM	1948
—	—	R/Hornsby (423661)	0-4-0DM	1958

Stock

3 ex-SECR 'Birdcage' coaches
1 ex-LSWR coach
5 4-wheel coaches of NLR, GER, SECR and LC&DR origins
1 LNWR Director's saloon

Passenger trains: Tenterden-Rolvenden: Sundays only January, February 11.15-15.00

Tenterden-Northiam: Sundays only March, November 11.00-16.00; Weekends 30 March-27 October 11.00-16.30; also Wednesdays and Thursdays June and September; Mondays-Fridays 29 March-7 April, 25 May-2 June, 1 July-1 September (11.00-16.00)
Special events: Please see press or phone for details
Facilities for disabled: A special coach for disabled people, 'Petros', is allocated to most trains (telephone for confirmation of availability), reserved parking at Tenterden.
Toilets with disabled access at Tenterden and Northiam, and in coach 'Petros'
Special notes: 'Wealden Pullman' dining car service operates on most Saturday evenings in season. Advance booking is essential for this train. Santa special services operate on each Saturday and Sunday in December. Advanced booking recommended
Membership details: New Members Secretary c/o above address
Feature article reference: RW August 1989, p460 (Northiam extension); RW October 1989, p629 (diesels)

1 GER observation car
SECR and LSWR Family saloons
3 Pullman cars
8 ex-BR Mk 1 coaches
1 ex-BR Mk 1 RU 'Diana'
9 ex-SR Maunsell coaches
3 steam cranes
Large interesting collection of freight vehicles, totalling 51 vehicles

Owners
10 (London Borough of Sutton)
21 and 22 (Kent & East Sussex Loco Trust)
19 (Norwegian Locomotive Trust)

Below:
The first passenger train since 1954 leaves Northiam station on the newly extended K&ESR on 18 May 1990. It was double-headed by Nos 10 *Sutton* and 25 *Northiam*. *Jim Berryman*

Lavender Line — SC

The neat appearance of the station and line at Isfield is largely due to the care lavished on it by its owner, David Milham and his family.
Location: Isfield Station, Isfield, near Uckfield, East Sussex TN22 5XB
OS reference: TQ 452171
Operating society/organisation: The Lavender Line Steam Museum
Telephone: Isfield (082 575) 515
Car park: On site
Access by public transport: Train to Uckfield or Lewes, then bus to Isfield. Buses run every two hours

Locomotives

Name	*No*	*Origin*	*Class*	*Type*	*Built*
—	15224	BR	12	0-6-0DE	1949

Industrial locomotives

Name	*No*	*Builder*	*Type*	*Built*
Annie	945	A/Barclay (945)	0-4-0ST	1904
—	68012	Hunslet (3193)	0-6-0ST	1944

Stock
4 ex-BR Mk 1 coaches
1 ex-GWR 'Toad' brake van

On site facilities: Refreshments, licensed bar, wine & dine trains (advanced booking essential), souvenir shop, museum
Length of line: 1½ miles
Public opening: Every Sunday and Bank Holiday Monday, March-November, 10.30-17.00
Facilities for disabled: Yes
Special events: Santa specials in December
Special notes: Luxury wine & dine trains every Friday, all year except Feb 19.30-20.00, booking essential

London Transport Museum M

The museum was opened in the converted Flower Market in March 1980. The displays cover nearly two centuries of public transport in London
Location: Covent Garden, London WC2E 7BB
OS reference: TQ 303809
Operating society/organisation: London Regional Transport
Telephone: 071-379 6344
Access by public transport: Underground to Covent Garden, Leicester Square or Charing Cross. Buses to Strand or Aldwych
On site facilities: Museum shop, lecture theatre, photo and research libraries (by appointment)
Public opening: Daily 10.00-18.00 (last admissions 17.15). Closed 24/25/26 December. Reduced admission prices for children, students, senior citizens, UB40 holders and pre-booked parties. Free admission for registered disabled
Facilities for disabled: Disabled toilets available, wheelchair access to most of the displays. Also taped guide for visually handicapped visitors. Free admission for registered disabled visitors and person accompanying them. Please advise in advance if a party of disabled visitors would like to visit
Special notes: Visitors can operate the controls of a tram, a bus, an Underground train, signals and points. In addition to the vehicles and rolling stock there are models, signs, posters, paintings, audio-visual displays and a 1906 Otis lift car. At certain times there will be films and temporary exhibitions. In addition to the vehicles on display there is a growing reserve collection
Membership details: There is a Friends of the London Transport Museum, please contact above address for details
Feature article reference: RW March 1989, p150

Locomotives

Name	*No*	*Origin*	*Class*	*Type*	*Built*
—	23	Met Rly	A	4-4-0T	1866
John Hampden	5	Met Rly		Bo-Bo	1922

Industrial locomotives

Origin	*Builder*	*Type*	*Built*
Wotton Tramway	A/Porter (807)	0-4-0TG	1872

Electric stock
4248 District Rly Q23 stock driving motor coach 1923
11182 LPTB 1938 stock driving motor coach
400 Met Rly bogie stock coach 1899
30 City & South London Rly 'Padded Cell' coach 1890
Great Northern Piccadilly & Brompton Railway 'Gate stock' car 1906
13 City & South London Railway locomotive of 1890, on loan from Science Museum for 'Tube Centenary' exhibition

Stock
1 Met Rly milk van
3 electric trams
2 horse buses
5 motor buses
1 trolleybus

Below:
The London Transport museum holds a number of interesting items, some of which can be seen in this view. *LT Museum*

North Downs Steam Railway

SC

The North Downs Steam Railway is a rapidly developing society, establishing a museum and working railway, on a greenfield site at Stone Lodge, Dartford
Location: Stone Lodge Centre, Dartford, Kent
Operating society: North Downs Steam Railway
Telephone: Dartford (0322) 28260
Main station: Cotton Lane — free car parking available
Other station: London Road — (new station to open during 1991)
Directions: By car — leave M25 Junction 1a (Dartford Tunnel). Follow Stone Lodge/Historic Dartford signs up Cotton Lane (2min).
By train — Network SouthEast to Dartford. Then Kentish Bus to Stone Lodge Farm entrance (routes 480 and 481), 5min walk
Length of line: ½-mile
Public opening: Every Sunday and bank holiday all year round. Saturdays from 30 March. Passenger trains hourly from Cotton Lane
Special events: 31 March/1 April — Friends of Thomas; 18/19 May — Industrial Weekend; 1/2 June — Gala Weekend; 31 August/1 September — Friends of Thomas; 15 September — Children's Day. Santa Trains every weekend in December

Industrial locomotives

Name	*No*	*Builder*	*Type*	*Built*
Scottie	1	R/Hornsby (412427)	4wDM	1957
Burt	2	M/Rail (9019)	4wDM	1951
North Downs	3	RSH (7846)	0-6-0T	1955
—	4	Fowler (4220008)	0-4-0DH	1959
Crabtree	5	R/Hornsby (338416)	4wDM	1953
Princess Margaret	6	Barclay (376)	0-4-0DM	1948
Telemon	7	Drewry/Vulcan (D295)	0-4-0DM	1955
Octane	8	YEC (2686)	0-4-0DE	1960
—	9	R/Hornsby (512572)	4wDM	1965
Topham	10	Bagnal (2193)	0-6-0ST	1922
Paxman	—	R/Hornsby (412718)	0-4-0DE	1958

Stock
2 ex-LT T stock DMB cars, 2758 and 2749
1 ex-LCDR 4-wheel carriage (body only)
1 ex SECR carriage (body only)
2 ex-BP oil tanks
1 ex-Blue Circle open cement wagon
1 ex-LT brake van chassis
1 ex-SR Bogie van
1 ex-SR PMV
1 ex-LT brake van
1 Grafton 5-ton diesel-hydraulic crane
1 ex-Esso oil tank

Facilities for disabled: Special compartment in train for wheelchairs and special toilet facilities
Membership and service details: North Downs Steam Railway, 29 Southbourne, Washford Farm, Ashford, Kent TN23 2UB
Tel: Ashford (0233) 46170

North Woolwich Old Station Museum

M

No expense has been spared in the very imaginative restoration of this attractive Victorian terminus building overlooking the Thames. Railway artefacts, documents, drawings, etc are well displayed in glass cases or on the walls, the stock being stabled in the platform area. Convenient for the new City airport and connections for the Docklands Railway
Location: North Woolwich Old Station Museum, Pier Road, North Woolwich, London E16 2JJ
OS reference: TQ 433798
Organisation: Passmore Edwards Museum Governors
Telephone: 071-474 7244
Car park: Only in adjoining streets
Public transport: BR North London Link. Buses: 101, 69 and 58

Locomotives

Name	*No*	*Origin*	*Class*	*Type*	*Built*
—	229	GER	209	0-4-0ST	1876

Industrial locomotives

Name	*No*	*Builder*	*Type*	*Built*
—	—	Hibberd (3294)	4wDM	1948
—	—	Peckett (2000)	0-6-0ST	1942
—	—	RSH (7667)	0-6-0ST	1950

Stock
1 ex-LNER coach
2 compartment sections of LTSR coach

Facilities: Museum shop
Public opening: Saturday to Wednesday, 10.00-17.00. Sunday and Bank Holidays, 14.00-17.00. Closed Thursday and Friday

Romney, Hythe & Dymchurch Railway

This line was built in 1926/27 as a one-third size miniature main line, and is by far the longest and most fully-equipped 15in gauge railway in the world. It carries not only daytrippers and holidaymakers but also children to and from the local school at New Romney.
Headquarters: Romney, Hythe & Dymchurch Railway, New Romney station, Kent TN28 8PL
Telephone: New Romney (0679) 62353/63256
OS reference: TR 074249
Main station: Hythe
Other public stations: Dymchurch, Jefferstone Lane, New Romney, Romney Sands, Dungeness
Car parks: Hythe, Dymchurch, New Romney, Dungeness
Access by public transport: Sandling Junction BR station (1½ miles), Folkestone Central BR station — 4 miles: bus connection from Folkestone bus station to Hythe
Refreshment facilities: New Romney, Dungeness. Also an observation coach on certain trains
Souvenir shops: Hythe, New Romney
Museum: New Romney
Depot: New Romney
Length of line: 13½ miles, 15in gauge
Passenger trains: Train frequency depends on the time of year: maximum frequency is 45 minutes
Period of public operation: Daily from Easter until end of September, and at weekends in March and October

Locomotives

Name	*No*	*Builder*	*Type*	*Built*
Green Goddess	1	Davey Paxman	4-6-2	1925
Northern Chief	2	Davey Paxman	4-6-2	1925
Southern Maid	3	Davey Paxman	4-6-2	1926
The Bug	4	Krauss (8378)	0-4-0TT	1926
Hercules	5	Davey Paxman	4-8-2	1926
Samson	6	Davey Paxman	4-8-2	1926
Typhoon	7	Davey Paxman	4-6-2	1926
Hurricane	8	Davey Paxman	4-6-2	1926
Winston Churchill	9	YEC (2294)	4-6-2	1931
Doctor Syn	10	YEC (2295)	4-6-2	1931
Black Prince	11	Krupp (1664)	4-6-2	1937
John Southland	12	TMA Birmingham	Bo-Bo	1983
—	14	TMA Birmingham	Bo-Bo	1989
—	PW1	M/Rail (7059)	4wDM	1938
—	PW2	RH&DR	4wPM	1965
Redgauntlet	PW3	RH&DR	4wPM	1975

Locomotive notes: Nos 6, 9 and 11 are expected to be on loan to the Bure Valley Railway during 1991

Stock

42 saloon bogie coaches
12 open bogie coaches
3 semi-open coaches
5 luggage/brake saloons
1 Parlour car
1 mess coach
40 assorted wagons

Special events: Enthusiasts' Days 19 May — Steam & Diesel Gala Day; 8 September — Military Railway Day
Special notes: Senior citizen concession Fridays, Saturdays, Sunday, (return journey for single fare). The 'Romney Toy and Model Museum' at New Romney. Parties can be catered for at New Romney
Membership details: RH&DR Association, 26 Norman Close, Battle, East Sussex TN33 0BD
Feature article reference: RW June 1990, p334

Below:
Storm time at New Romney as *Hurricane* and *Typhoon* are coaled prior to the commencement of the day's service. *Peter Groom*

Ruislip Lido Railway

SC

The 12in gauge line is operated by enthusiast volunteers as an attraction within Ruislip Lido, a country park which is maintained by the London Borough of Hillingdon
Location: Ruislip Lido, Reservoir Road, Ruislip, Middlesex
Operating society/organisation: Ruislip Lido Railway Society Ltd, Membership Secretary, Mrs S. Simmons, 9 Wiltshire Lane, Eastcote, Pinner, Middx HA5 2LH
Telephone: 081-866 9654
Car park: Available at Lido
Access by public transport: Ruislip Underground station (Metropolitan and Piccadilly lines) then by bus H13 (Monday-Saturday), 114 (Sundays) or E2 (Summer Sundays). Lido is off the A4180 road
Refreshment facilities: Available at Lido site
Length of line: 1-mile single journey, 2 miles return with a planned extension of about 1 mile which will terminate beside main entrance to car park
Public opening: The line is open Sundays throughout the year, Monday, Wednesday, Friday and Saturday in April, May, September and October. Every day in June, July and August, also daily during all school holidays
Journey time: Single 12min, return 22min
Facilities for disabled: Limited number of wheelchair passengers can be accepted for travel
Membership details: c/o above address

Locomotives

Name	*No*	*Builder*	*Type*	*Built*
Robert	3	Severn-Lamb	B-2 PH	1973
Lady of the Lakes	5	Ravenglass & Eskdale Railway	B-B DH	1985
Graham Alexander	7	Severn-Lamb	B-B DH	1990

Locomotive notes: All locomotives are available for service and a steam locomotive is under construction

Stock
4 open coaches, with a further six under construction
5 closed coaches
Miscellaneous service stock

Below:
Ruislip Lido Railway Society are currently building a steam locomotive, in the meantime diesels operate the service. *Lady of the Lakes* and *Graham Alexander* are seen here on 26 March 1990.

Science Museum

M

Built on land acquired with the profits from the Great Exhibition of 1851, the Science Museum was one of the first to include industrial archaeology. Here you will find the originals featured in the history books, such as Robert Stephenson's *Rocket*.
Location: South Kensington
OS reference: TQ 268793
Operating society/organisation: Science Museum, Exhibition Road, South Kensington, London SW7
Telephone: 071-938 8000

Locomotives

Name	*No*	*Origin*	*Class*	*Type*	*Built*
Caerphilly Castle	4073	GWR	'Castle'	4-6-0	1923
Deltic	—	E/Electric	—	Co-Co	1955
Rocket	—	L&MR	—	0-2-2	1829
Rocket	—	(Replica)	—	0-2-2	1935
Sans Pareil	—	L&MR	—	0-4-0	1829
	13	C&SLR	—	4wE	1890
Puffing Billy	—	Wylam Colliery	—	0-4-0	1813

Industrial locomotives

Name	*No*	*Builder*	*Type*	*Built*
Bauxite	2	B/Hawthorn (305)	0-4-0ST	1874

Access by public transport: South Kensington Underground station
Catering facilities: Cafe on 3rd floor, tea, coffee, sandwiches etc
On site facilities: Book shop
Public opening: Weekdays 10.00-18.00, Sundays 11.00-18.00. Closed 1 January, Good Friday, May Day Monday, 24-26 December
Special events: All organised by the National Railway Museum, York, which is part of the Science Museum. Telephone (0904) 621261 for details
Facilities for disabled: Toilets on ground floor, ramp and lifts to all floors (except basement). Parties should contact before arrival if extra assistance is required
Special notes: Static exhibits only

Stock
LPTB Underground coach, 3327 (built 1927)

Locomotive notes: All restored to static display condition

Sittingbourne & Kemsley Light Railway SC

The Sittingbourne & Kemsley Light Railway is part of the 2ft 6in gauge railway built to convey paper and other materials between mills at Sittingbourne and Kemsley and the Dock at Ridham on the banks of the Swale. The first section of the line opened in 1906 and two of the engines then in use remain on the line today.

The railway is now leased from U.K. Paper Group and is operated as a tourist attraction. Passenger trains are normally steam-hauled and are formed of a varied selection of open and covered coaches. For the first half mile of the journey the train twists and turns through Milton Regis on a

Locomotives

Name	*No*	*Builder*	*Type*	*Built*
Alpha	—	Bagnall (2472)	0-6-2T	1932
Triumph	—	Bagnall (2511)	0-6-2T	1934
Superb	—	Bagnall (2624)	0-6-2T	1940
Unique	—	Bagnall (2216)	2-4-0F	1924
Premier	—	K/Stuart (886)	0-4-2ST	1905
Leader	—	K/Stuart (926)	0-4-2ST	1905
Melior	—	K/Stuart (4219)	0-4-2ST	1924
Edward Lloyd	—	R/Hornsby (435403)	4wDM	1961
Victor	—	Hunslet (4182)	4wDM	1953

Below:
Superb **approaches Burley crossing with an afternoon service from Sittingbourne.** *D. C. Green*

concrete viaduct which was one of the first reinforced concrete structures to be built
Headquarters: Sittingbourne & Kemsley Light Railway Ltd, The Wall, Milton Regis, Sittingbourne, Kent
Telephone: Sittingbourne (0795) 424899 (talking timetable) or (0634) 52672 (other enquiries)
Main station: Sittingbourne
Car park: Sittingbourne
Access by public transport: Sittingbourne BR station, A2 and M2 roads
OS reference: Sittingbourne 905643, Kemsley Down 920662
Refreshment facilities: Kemsley Down
Souvenir shop: Sittingbourne
Depot: Kemsley Down (access by rail only)
Length of line: 2 miles, 2ft 6in gauge
Passenger trains: Ex-industrial line Sittingbourne-Kemsley Down
Journey time: 15min each way

Industrial standard gauge locomotives

Name	*No*	*Builder*	*Type*	*Built*
—	4	H/Leslie (3718)	0-4-0ST	1928
Bear	—	Peckett (614)	0-4-0ST	1896
—	1	Barclay (1876)	0-4-0F	1925

Locomotive notes: In service: *Triumph*, *Superb*, *Premier*. Under repair: *Leader*, *Melior*, *Victor*. On static display *Alpha*, *Unique* and standard gauge exhibits

Stock
10 bogie coaches (4 ex-Chattenden & Upnor Railway)
2 open coaches
Various wagons

Period of public operation: Good Friday, Easter Monday and bank holidays; Saturdays 30 March; 4, 25 May, and during August; Sundays 31 March-mid October; Wednesdays 29 May and 7-28 August. Santa specials, contact for details
Special events: Special open day — 13 October
Special notes: There is no public access to Kemsley Down other than by the railway on operating dates. When the line is closed all stock is stored in security compounds, on the mill premises. Family ticket available, special rates for parties and senior citizens
Membership details: Mrs Stickler, Lite Hjem, Woodlands Estate, Blean, Canterbury, Kent CT2 9JN

Southall Railway Centre SC

During the autumn of 1988 the GWRPG moved all their stock into the former BR diesel depot adjacent to their old site. At the time of writing the area was being up-graded to allow the general public into the site, this involves fencing the centre off from the adjacent BR tracks. Please see press for details of open days etc
Location: The former BR diesel depot, approximately 10min from BR station. Access is only possible via the footbridge at the London end of the station, during opening times only
Operating society/organisation: GWR Preservation Group
Telephone: 081-574-1529 R. A. Gorringe (evening and weekend only)
Facilities for disabled: Access to the site is not possible, involving two flights of steps, those wishing to visit are advised to make contact before their visit and alternative arrangements will be made

Locomotives

Name	*No*	*Origin*	*Class*	*Type*	*Built*
—	2885	GWR	2884	2-8-0	1938
—	4110	GWR	5101	2-6-2T	1936
—	9682	GWR	5700	0-6-0PT	1949
—	68078*	LNER	J94	0-6-0ST	1946

*Not on site

Industrial locomotives

Name	*No*	*Builder*	*Type*	*Built*
—	—	AEC Southall	4wDM	1938
—	2100	Peckett (2100)	0-4-0ST	1949
Birkenhead	7386	RSH (7386)	0-4-0ST	1948
Francis Baily of Thatcham	AD251	R/Hornsby (390772)	0-4-0DM	1957

Stock
1 BR 'Gane' wagon
1 BR 'Rectank' wagon
1 BR Generator van
1 BR Goods van
1 GWR 'Mink' van
1 GWR brake van
2 BP oil tank wagons
1 GWR bogie bolster wagon
1 LMS parcels van
1 LMS stores van
1 Staff coach, former LMS BCK
1 Wagons Lits coach

Owners
2100 (the City of Portsmouth Museum)

Central Southern

Buckinghamshire Railway Centre SC

Formerly known as Quainton Road, the name of the station, the Centre has been improved considerably during recent years. Its true potential could, however, be realised by an enlightened approach allowing operation of steam trains to and from Marylebone

Location: Adjacent to BR goods only line to Aylesbury. Turn off A41 at Waddesdon 6 miles NW of Aylesbury, Bucks

OS reference: SP 738190

Operating society/organisation: Quainton Railway Society Ltd, The Railway Station, Quainton, near Aylesbury, Bucks HP22 4BY

Telephone: Quainton (029675) 450

Car park: Quainton Road — Free parking

Access by public transport: BR Aylesbury station. Special bus on Bank Holidays connects Centre with BR trains

On site facilities: Souvenir bookshop, light refreshments, toilets, steam-hauled train rides. Museum of small relics, secondhand bookshop, miniature railway

Catering facilities: Victorian Tea specials. Afternoon tea in LNWR dining car, write for details to catering manager. Lunches and afternoon teas by arrangement

Length of line: Two ½-mile demonstration lines

Locomotives

Name	*No*	*Origin*	*Class*	*Type*	*Built*
—	1	Met Rly	E	0-4-4T	1898
—	0314	LSWR	0298	2-4-0WT	1874
Wightwick Hall	6989	GWR	'Hall'	4-6-0	1948
—	7200	GWR	7200	2-8-2T	1934
—	7715	GWR	5700	0-6-0PT	1930
—	9466	GWR	9400	0-6-0PT	1952
—	41298	LMS	2MT	2-6-2T	1951
—	41313	LMS	2MT	2-6-2T	1952
—	46447	LMS	2MT	2-6-0	1950
—	D2298	BR	04	0-6-0DM	1960
—	25057	BR	25	Bo-Bo	1963

Industrial locomotives

Name	*No*	*Builder*	*Type*	*Built*
Sydenham	—	A/Porter (3567)	4wTG	1895
Scott	—	Bagnall (2469)	0-4-0ST	1932
—	—	Baguley (2161)	0-4-0DM	1941
Swanscombe	—	Barclay (699)	0-4-0ST	1891
—	GF3	Barclay (1477)	0-4-0F	1916
Tom Parry	—	Barclay (2015)	0-4-0ST	1935
—	—	Barclay (2243)	0-4-0F	1948
Osram	—	Fowler (20067)	0-4-0DM	1933
—	3	H/Leslie (3717)	0-4-0ST	1928
Sir Thomas	—	H/Clarke (1334)	0-6-0T	1918
—	—	H/Clarke (1742)	0-4-0ST	1946

Right:
An unusual 'one-off' run over part of London's underground was organised by LTL for a week-day evening steam-hauled journey over the Stanmore line. The locomotive used was the Buckinghamshire Railway Centre's Metropolitan 0-4-4T No 1, which is seen on arrival at Stanmore, 26 July 1990.
Hugh Ballantyne

Public opening: Sundays and Bank Holidays: Easter-end October. Wednesdays in June, July and August. Limited opening (no engines in steam) Saturdays, June to September
Special events: Santa Specials in December (please write for details), others throughout the year. Write for details
Facilities for disabled: Access to most of site, but not toilets. No advance notice required
Special notes: One of the largest collection of standard gauge locomotives, together with a most interesting collection of vintage coaching stock, much of which was built in the last century

—	—	Hunslet (2067)	0-4-0DM	1940
Arthur	—	Hunslet (3782)	0-6-0ST	1953
Juno	—	Hunslet (3850)	0-6-0ST	1958
—	65	Hunslet (3889)	0-6-0ST	1964
—	66	Hunslet (3890)	0-6-0ST	1964
—	26	Hunslet (7016)	0-6-0DH	1971
Redland	—	K/Stuart (K4428)	0-4-0DM	1929
Coventry No 1	—	NBL (24564)	0-6-0ST	1939
—	—	Peckett (1900)	0-4-0T	1936
Gibraltar	—	Peckett (2087)	0-4-0ST	1948
—	—	Peckett (2104)	0-4-0ST	1948
—	—	Peckett (2105)	0-4-0ST	1948
—	T1	Hibberd (2102)	4wD	1937
Tarmac	—	Hibberd (3765)	0-4-0DM	1955
—	—	Sentinel (6515)	4wVBTG	1926
—	11	Sentinel (9366)	4wVBTG	1945
—	7	Sentinel (9376)	4wVBTG	1947
—	—	Sentinel (9537)	4wVBTG	1947
Chislet	9	Yorkshire (2498)	0-6-0ST	1951

Stock
1 LCDR 1st Class 4 wheeler
1 MSLR 3rd Class 6 wheeler
4 LNWR coach bodies
2 GNR 6 wheelers
3 LNWR coaches
3 LMSR coaches
1 BR(W) Hawksworth brake 3rd
2 BR Mk 1 coaches
1 BR Mk 2 coach
1 BR Suburban brake
3 LNER coaches
1 LNWR full brake 6 wheeler
1 LNWR combination truck
1 LSWR ventilated fruit van
1 SR PMV
1 LMSR passenger brake van
1 GWR passenger brake van
1 BR(W) 'Siphon G'
1 BR horse box
1 BR CCT

3 ex-London Underground coaches
1 2ft gauge post office mailbag car 803
Sentinel/Cammell 3-car steam railcar unit 5208 (ex-Egyptian National)
Numerous goods vehicles/wagons/vans

Owners
41298, 41313 and 46447, *Juno* (the Ivatt Locomotive Trust)
9466 (the 9466 Group)

Cholsey & Wallingford Railway — SC

Location: Hithercroft Industrial Estate, Wallingford, Oxfordshire
Operating Society: Cholsey & Wallingford Railway Preservation Society, PO Box 16, Wallingford, Oxon OX10 0NF
Telephone: 0491 35067 (weekends and Bank Holidays only)
Car park: Habitat adjacent to Wallingford station
Access by public transport: South Midland Bus Company, BR Didcot Parkway or Cholsey
Public Opening: Every Sunday Easter-end September 11.00-17.00. Trains run first and last Sunday and bank holidays 13.00-16.30
Length of line: 1¼ miles
Journey time: Approximately 20min

Locomotives

Name	*No*	*Origin*	*Class*	*Type*	*Built*
—	4247	GWR	4200	2-8-0T	1916
George Mason	08 123	BR	08	0-6-0	1955

Industrial locomotives

Name	*No*	*Builder*	*Type*	*Built*
Walrus	3271	Planet (3271)	0-4-0	1949
—	304470	R/Hornsby	0-4-0	1951
Thames	3	Barclay (2315)	0-4-0ST	1951

On site facilities: Souvenir and coffee shops. Museum, model railway. Children's mini railway
Special events: 31 March/1 April — Easter Bunnies; 5/6 May — Spring Bank Holiday Fair; 10 August — Cholsey Festival; 25/26 August — Friends of Thomas; 8, 15, 22 December — Santa Specials
Special notes: Light Railway Order now granted and work is proceeding to return to Cholsey (BR) bay platform
Membership details: Bob Morrison, at above address

Didcot Railway Centre SC

Based around the GWR engine shed and dcpot, thc Centre now has a typical GWR small country station with signalbox (from Radstock), re-creation of Brunel's broad gauge railway, two demonstration lines, and a small relics museum
Location: Adjacent to BR station, Didcot, Oxfordshire. Access via station subway
OS reference: SU 525907
Operating society/organisation: Great Western Society Ltd, Didcot Railway Centre, Didcot, Oxon OX11 7NJ
Telephone: Didcot (0235) 817200
Car park: Didcot BR station
Access by public transport: Didcot Parkway BR station
Refreshment facilities: Refreshment room open all days centre is open (lunches, snacks)
On site facilities: GWR locomotive depot, replica GWR station, museum and broad gauge demonstration. Souvenir sales. Rides are available on the demonstration lines on steamdays
Length of line: 1,000yd
Public opening: Saturdays and Sundays all year. Daily 29 March-7 April and 25 May-15 September. Steamdays first and last Sunday each month from March, Bank Holidays, all Sundays June-August and all Wednesdays in August. Open 11.00-17.00
Train rides: On Steamdays there is normally continuous operation of the passenger train, interrupted by freight train demonstrations and turning of the locomotives on some days
Special events: Spring Steam Gala 25 May-27 May; Autumn Steam Gala 28/29 September; Photographers Evening 26 October; Santa Steamings 1, 8, 15, 22 December; New Year Steamings 29 December, 1 January 1992
Facilities for disabled: Steps at access from BR subway may cause problems but assistance can normally be provided (advance notification is useful)
Membership details: Brian Phillips, at above address
Feature article references: RW June 1989, p333 (No 6024); RW December 1989, p717 (No 71000); RW June 1990, p368 (No 71000)

Locomotives

Name	*No*	*Origin*	*Class/builder*	*Type*	*Built*
—	22	GWR	Diesel Railcar	1A-A1	1940
—	1338	GWR Cardiff Rly	Kitson (3799)	0-4-0ST	1898
Trojan	1340	GWR	Avonside (1380)	0-4-0ST	1897
—	1363	GWR	1361	0-6-0ST	1910
—	1466	GWR	1400	0-4-2T	1936
—	3650	GWR	5700	0-6-0PT	1939
—	3738	GWR	5700	0-6-0PT	1937
—	3822	GWR	2884	2-8-0	1940
—	4144	GWR	5101	2-6-2T	1946
Maindy Hall	4942	GWR	'Hall'	4-6-0	1929
Nunney Castle	5029	GWR	'Castle'	4-6-0	1934
Drysllwyn Castle/ Earl Bathurst	5051	GWR	'Castle'	4-6-0	1936
—	5322	GWR	4300	2-6-0	1917
—	5572	GWR	4575	2-6-2T	1927
Hinderton Hall	5900	GWR	'Hall'	4-6-0	1931
King Edward II	6023	GWR	'King'	4-6-0	1930
King Edward I	6024	GWR	'King'	4-6-0	1930
—	6106	GWR	6100	2-6-2T	1931
—	6697	GWR	5600	0-6-2T	1928
Burton Agnes Hall	6998	GWR	'Hall'	4-6-0	1949
—	7202	GWR	7200	2-8-2T	1934
Cookham Manor	7808	GWR	'Manor'	4-6-0	1938

Below:
The ARPS/Ian Allan Railway Heritage award for the best restored signalbox in 1989 was given to the GWS's Bristol Group for Radstock North box, now at Didcot. The box is at the centre's branch terminus station. *L. Waters/GWS*

Duke of Gloucester	71000	BR	8P	4-6-2	1954
—	D7018	BR	35	B-B	1962
Pontyberem	2	Burry Port & Gwendraeth Valley Rly		0-6-0ST	1900
Shannon	5	Wantage Tramway		0-4-0WT	1857

Industrial locomotives

Name	*No*	*Builder*	*Type*	*Built*
Bonnie Prince Charlie	1	RSH (7544)	0-4-0ST	1949
—	26	Hunslet (5238)	0-6-0DH	1962

Locomotive notes: Locomotives available in 1990 should be: 3822, 5029, 5572, 6024, 6106, 6998 and D7018. Locomotives under restoration include: 1466, 3650, 7202. Construction of the Firefly Trust's reproduction broad gauge locomotive *Firefly* is being undertaken

Stock

Over 40 ex-GWR coaches are preserved along with numerous ex-GWR freight wagons

Owners

5 (on loan from the National Railway Museum)
D7018 (Diesel & Electric Group)
6024 (the 6024 Preservation Society)
71000 (the Duke of Gloucester Steam Locomotive Trust)

Gloucestershire Warwickshire Railway

Part of an ambitious project to link Cheltenham racecourse with Stratford racecourse, much has been done to recreate the railway and buildings that made up this cross-country route. The railway is home to many owners of private locomotives and rolling stock, so from time to time the items on display may vary

Location: Toddington Station, Toddington
OS reference: SO 050322
Operating society/organisation: Gloucestershire Warwickshire Steam Railway PLC, The Station, Toddington, Cheltenham, Glos GL54 5DT
Telephone: Toddington (0242) 621405
Main station: Toddington
Other public stations: Winchcombe, Gretton Meadow
Access by public transport: Public transport is very limited with occasional buses from Cheltenham, Stratford on Avon and Evesham only. Local bus service Castleways will answer timetable queries on (0242) 602949
Car park: On site
On site facilities: Sales, catering, museum, narrow gauge rides, toilets
Length of line: 4 miles
Public opening: During the week the station is unmanned but visitors are

Above:
The NRM's No 3440 *City of Truro* was loaned to the GWR for their Gala Weekend of 13/14 October 1990. It is seen on the Sunday steaming through Winchcombe station. *Stephen Widdowson*

Locomotives

Name	*No*	*Origin*	*Class*	*Type*	*Built*
—	2807*	GWR	2800	2-8-0	1905
—	4277	GWR	4200	2-8-0T	1920
Kinlet Hall	4936	GWR	'Hall'	4-6-0	1929
Defiant†	5080	GWR	'Castle'	4-6-0	1939
—†	7752	GWR	5700	0-6-0PT	1930
Peninsular and Oriental SNCo	35006	SR	MN	4-6-2	1941

welcome. Public services on Sundays, Bank Holiday Mondays, plus some Saturdays between March and October
Special events: Santa Specials in December; Mince Pie Specials and New Year specials on 1 January 1992
Special notes: The site is being developed as the headquarters of the railway between Cheltenham and Stratford. The GWR owns the railway land between Cheltenham and Honeybourne, and has been operating over 2½-miles from Toddington to Winchcombe. 1988 saw the track relaid all the way to Gretton and trains started running there from Easter 1990
Feature article reference: RW January 1991, p29 (Portrait of . . .)

—	76077	BR	4MT	2-6-0	1956
—	D9537	BR	14	0-6-0DH	1965
—	D9539	BR	14	0-6-0DH	1965
—	D9553	BR	14	0-6-0DH	1965

*Transferred to Birmingham Railway Museum, Tyseley for restoration
†On loan from Birmingham Railway Museum, Tyseley

Industrial locomotives

Name	*No*	*Builder*	*Type*	*Built*
Huntsman	—	Bagnall (2655)	0-6-0ST	1941
John	—	Peckett (1976)	0-4-0ST	1939
Robert Nelson No 4	—	Hunslet (1800)	0-6-0ST	1936
King George	—	Hunslet (2409)	0-6-0ST	1942
—	19	Fowler (4240016)	0-6-0DH	1964
—	21	Fowler (4210130)	0-4-0DM	1957
—	—	H/Clarke (D615)	0-6-0DM	1938
—	—	Hibberd (2893)	4wPM	1943

Stock
3 ex-GWR coaches
13 ex-BR coaches
1 ex-LMS coach
Numerous plus wagons

Owners
2807 and 4277 (the Cotswold Steam Preservation Ltd)
35006, 76077 (the P and O Locomotive Society)
D9537, D9539 and D9553 (Cotswold Diesel Preservation Group)

Industrial narrow gauge locomotives (2ft gauge)

Name	*No*	*Builder*	*Type*	*Built*
Isibutu	5	Bagnall (2820)	4-4-0T	1946
George B	—	Hunslet (680)	0-4-0ST	1898
Chaka	—	Hunslet (2075)	0-4-2T	1940
Justine	—	Jung (939)	0-4-0WT	1906
—	—	Henschel (15968)	0-8-0T	1918
—	2	Lister (34523)	4wDM	1949
—	3	M/Rail (4565)	4wPM	1928
Spitfire	—	M/Rail (7053)	4wPM	1937
—	1	R/Hornsby (166010)	4wDM	1932
—	L5	R/Hornsby (181820)	4wDM	1936
—	—	R/Hornsby (354028)	4wDM	1953

Stock
1 coach
11 wagons

Hampshire Narrow Gauge Railway Society SC

Location: Close to Southampton and Winchester, on private property. The site is only open to the public on advertised dates, other times by application (with SAE) to address below
OS reference: SU 522173
Operating society/organisation: Hampshire Narrow Gauge Railway Society, c/o 4 Holmdale Road, Gosport, Hants PO12 4PJ

Locomotives

Name	*No*	*Builder*	*Type*	*Built*
Wendy	1	Bagnall (2091)	0-4-0ST	1919
AGW1 Pet	2	M/Rail (4724)	4wP	1939
Cloister	3	Hunslet (542)	0-4-0ST	1891
Brambridge Hall	4	M/Rail (5226)	4wP	1936
—	5	R/Hornsby (392117)	4wD	1956
Josephine	6	Hunslet (1842)	0-4-2ST	1936
—	7	Hanomag (8310)	0-8-0T	1918
—	8	O/Koppel (4013)	4wD	1933
Norden	9	O/Koppel (20777)	0-4-0D	1936

Car park: On site but very limited
On site facilities: 2ft gauge line, light refreshments and souvenir shop. Toilets (not yet suitable for disabled). Visitors can visit workshops on request
Public opening: Please send SAE to Secretary

Stock
1 bogie coach
Various freight stock
Carriage portion of ex-LSWR 4-2-4T *The Bug* (standard gauge)

Hollycombe Steam Collection SC

An unusual collection of traction engines, Bioscope, organs, steam gallopers, Mighty Emperor Burrell Steam Engine, sawmill and paddle steamer engine which also includes a 7¼in gauge steam railway which climbs some fair gradients through the woodlands and gardens
Location: Iron Hill, Hollycombe, near Liphook, Hants
OS reference: SU 852295
Operating society/organisation: Hollycombe Steam & Woodland Garden Society, Iron Hill, Liphook, Hants GU30 7LP
Telephone: Liphook (0428) 724900 (opening days only); otherwise (04203) 4740

Industrial locomotives

Name	*No*	*Builder*	*Type*	*Built*
Excelsior	—	A/Porter (1607)	2-2-0WTG	1880†
Caledonia	70	Barclay (1995)	0-4-0WT	1931*
Jerry M	38	Hunslet (638)	0-4-0ST	1895*
Newcastle	—	H/Leslie (2450)	0-4-0ST	1899
—	16	R/Hornsby	4wDM	1941*

*2ft gauge
†3ft gauge

Car park: On site
Access by public transport: Liphook BR station (1½ miles)
On site facilities: Shop and refreshments.
Public opening: Please see press for details
Special events: Please see press for details

Isle of Wight Steam Railway

Separated from the mainland by the Solent, the line's isolation encouraged the maintenance and retention of Victorian locomotives and coaching stock which still operate the line today. Its rural charm enhances its attraction for the island's holidaymakers during the summer season
Headquarters: Isle of Wight Steam Railway, Haven Street station, Ryde, Isle of Wight PO33 4DS
Telephone: Station: Isle of Wight (0983) 882204
Main station: Haven Street
OS reference: SZ 556898
Other public stations: Wootton and Ashey
Car park: Haven Street
Access by public transport: Service 1a bus from Ryde or Newport to Wootton, also seasonal services 3 and 43 to Havenstreet
Refreshment facilities: Light refreshments available
Souvenir shop: Haven Street

Locomotives

Name	*No*	*Origin*	*Class*	*Type*	*Built*
Freshwater	W8 (32646)	LBSCR	A1X	0-6-0T	1876
—*	11 (32640)	LBSCR	A1X	0-6-0T	1878
Calbourne	W24	LSWR	02	0-4-4T	1891
—	D2554	BR	05	0-6-0DM	1956
—	D2059	BR	03	0-6-0DM	1959

*depicted as IWCR No 11

Industrial locomotives

Name	*No*	*Builder*	*Type*	*Built*
Invincible	37	H/Leslie (3135)	0-4-0ST	1915
Ajax	38	Barclay (1605)	0-6-0T	1918
Tiger	—	N/British (27415)	0-4-0DH	1954

Locomotive notes: *Ajax* is not on public display.

Stock
4 IWR coaches (bodies only)
1 IWR coach
3 LBSCR coaches
3 SECR coaches
1 LCDR coach
5 LCDR coaches (bodies only)
1 LBSCR coach (body only)
1 crane
1 ex-BR ballast tamper
1 Wickham trolley
30 wagons
6 parcels vans
2 ex-LT hoppers
1 ex-BR Lowmac

Museum: Small exhibits museum at Haven Street
Depot: Haven Street
Length of line: 2 miles (5 miles on extension of services, 1991)
Passenger trains: 21 March-18 July Wootton-Havenstreet only; 21 July-end October Wootton-Smallbrook Jct
Period of public operation: 28 March-28 April — Thursdays, Sundays and bank holidays, also 2/3 April; May — Wednesdays, Thursdays and Sundays, also bank holidays and Tuesdays 7/28 May; 2 June-18 July — Tuesdays, Wednesdays, Thursdays and Sundays; 21 July-8 September — daily (except Saturdays); 11-29 September — Wednesdays, Thursdays and Sundays; October — Thursdays and Sundays
Special events: Summer Steam Extravaganza, 22-25 August; Santa Specials, in December until Xmas (please write for details)
Facilities for disabled: Limited facilities but can be catered for singly, or in groups (by prior arrangement), toilets available
Special notes: Summer Steam Show, August Bank Holiday Friday, Saturday, Sunday and Monday. Traction engines, fairground organs, vintage cars etc.
Membership details: Mr J. Price at above address
Feature article reference: RW October 1989, p621 (including BR)

Above:
A scene that could almost be out of the history books, except that No 11 carries Isle of Wight Central Railway livery in this Southern scene. The 'Terrier' looks most at home on 2 August 1990 as it runs into Wootton.
Peter Tandy

Mid-Hants Railway

The Mid-Hants line was often used by the Bournemouth expresses on diversions during steam days and was therefore built to main line standards. This has enabled the line's new owners to operate large and powerful locomotives over the line which now has a direct connection with BR
Headquarters: Mid-Hants Railway PLC, Alresford station, Alresford, Hants SO24 9JG
Telephone: Alresford (0962) 733810/734200
Main station: Alresford
Other public stations: Ropley, Medstead & Four Marks, Alton
OS reference: Alresford SU 588325, Ropley SU 629324
Car park: Alresford, BR car park Alton
Access by public transport: *BR train services* — just over 1hr from London. Through ticketing

Locomotives

Name	*No*	*Origin*	*Class*	*Type*	*Built*
—	120	LSWR	T9	4-4-0	1899
—	30499	LSWR	S15	4-6-0	1920
—	506	LSWR	S15	4-6-0	1920
—	31625	SR	U	2-6-0	1929
—	31806	SR	U	2-6-0	1926
—	31874	SR	N	2-6-0	1925
Bodmin	34016	SR	WC	4-6-2	1945
Tangmere	34067	SR	BB	4-6-2	1947
Swanage	34105	SR	WC	4-6-2	1950
British India Line	35018	SR	MN	4-6-2	1945
—	73096	BR	5MT	4-6-0	1956
—	76017	BR	4MT	2-6-0	1954
Sturdee	601	MoS	WD	2-10-0	1943
Franklin D. Roosevelt	701	USATC	S160	2-8-0	1944
—	D3358	BR	08	0-6-0DE	1957
—	D5217	BR	25	Bo-Bo	1961
—	D5353	BR	27	Bo-Bo	1961

arrangements available from Waterloo and all BR stations. Alternatively, travel to Winchester station and take the bus to Alresford.
Bus services — are operated by Alder Valley; telephone Guildford 575226 or Aldershot 23322 for details.
Refreshment facilities: Buffet service on most trains; 'West Country' buffet at Alresford; tea/coffee available at Alton when information office open
Catering facilities: The 'Watercress Belle' operates on the 1st and 3rd Saturdays April-October. Early booking is essential, please telephone to confirm seat availability
Souvenir shops: Alresford and Alton. Collector's corner at Ropley
Depot: Ropley
Length of line: 10 miles
Passenger trains: Alton-Alresford. Sundays 6 January-27 October; Saturdays and Bank Holidays 30 March-26 October; Tuesdays, Wednesdays and Thursdays 28 May-18 July; Daily 22 July-6 September. Santa Specials 23/24, 30 November, 1, 7/8, 14/15, 19-24 December (bookings commence end of August)
Journey time: Round trip 1hr 40min max
Special events: 10 February — Valentines Lunch; 28 April — Morris Day; 11/12 May — Spring Enthusiasts' Weekend; 8/9 June — Father's Day Specials; 23 June — 1960s Summer Holiday; 6/7 July — Steam & Transport Weekend; 21 July — Austin Day; 8 September — Bus Bonanza; 15 September — War on the Line; 6 October — Teddy Bears Day; 19/20 October — Autumn Enthusiasts' Weekend
Facilities for disabled: Toilets at Ropley, special facilities for disabled passengers on selected trains
Membership details: Membership Secretary, c/o above address
Feature article reference: RW July 1990, p402 (No 701)

Industrial locomotives

Name	*No*	*Builder*	*Type*	*Built*
—	4	Fowler (22889)	0-4-0DM	1939
Errol Lonsdale	196	Hunslet (3796)	0-6-0ST	1953

Stock
2 steam cranes
1 Plasser & Theurer AL250 lining machine
36 ex-BR Mk 1 coaches
1 ex-BR Mk 2 coach
8 ex-SR coaches
3 ex-LSWR coaches
Numerous goods vehicles

Owners
30499 and 506 (the Urie Locomotive Society)
34105 (the 34105 Light Pacific Group)
76017 (the Standard 4 Locomotive Group)

Below:
Bearing the 'Cunarder' headboard and the correct discs to represent a working from Southampton Docks to Waterloo, rebuilt 'West Country' No 34016 *Bodmin* nears Chawton with an Alton-bound train on 1 July 1990. *John H. Bird*

Swindon & Cricklade Railway SC

This is the only preserved section of the former Midland & South Western Junction Railway, the society having had to relay track and associated works
Location: Tadpole Lane, Blunsdon (approximately mid way between Blunsdon St Andrew and Purton)
Operating society/organisation: Swindon & Cricklade Railway, Blunsdon Station, Blunsdon, Swindon, Wiltshire SN2 4DZ
Telephone: Swindon (0793) 771615 (weekends only)
Station: Blunsdon
OS reference: SU 110897
Length of line: 1-mile
Car park: Tadpole Lane, Blunsdon
Refreshment facilities: Blunsdon station amenities building
Souvenir shop: Blunsdon station amenities building. Various sales stands on Open Days around station area. Museum in converted coach
Depot: Blunsdon
Public opening: Saturdays and Sundays throughout the year.
Passenger trains: Passenger services operate 1 January; 17, 30/31 March; 1 April; 26/27 May; 8, 29 September; 31 October (evening). Details from Public Relations Manager c/o above address
Special events: Santa Specials 7/8, 14/15, 21/22 December
Facilities for disabled: Access to shop and refreshments
Membership details: Membership Secretary c/o above address

Locomotives

Name	*No*	*Origin*	*Class*	*Type*	*Built*
Foremarke Hall	7903	GWR	'Hall'	4-6-0	1949
—	5637	GWR	56xx	0-6-2T	1924
—	1506 (08114)	BR	08	0-6-0DE	1955
—	D5022	BR	25	Bo-Bo	1965

Industrial locomotives

Name	*No*	*Builder*	*Type*	*Built*
—	—	Fowler (4210137)	0-4-0DM	1958
Woodbine	—	Fowler (21442)	0-4-0DM	1936
Richard Trevithick	—	Barclay (2354)	0-4-0ST	1955
Merlin/Myrddin	1371	Peckett (1967)	0-4-0ST	1939

Stock
5 BR Mk 1 coaches
1 BR DMU vehicle
2 GWR coaches
Selection of goods rolling stock
Wickham railcar

Owners
7903 (the Foremarke Hall Locomotive Group)
5637 (the 5637 Locomotive Group)

Swindon GWR Museum M

The Great Western Railway Museum began life as a lodging house for railway workers, and was then a Wesleyan Methodist Chapel for 90 years before being acquired by Swindon Borough Council and converted into a museum. Alongside the museum, in the heart of the railway village, visitors can glimpse the living conditions of the railworkers at the turn of the century, in the carefully restored workman's house
Location: Faringdon Road, Swindon, Wiltshire
OS reference: SU 145846
Operating society/organisation: Borough of Thamesdown, GWR Museum, Faringdon Road, Swindon, Wiltshire
Telephone: Swindon (0793) 526161 (ext 4552 or (0793) 493189
Car park: Swindon BR station and street parking
Access by public transport: Swindon BR (10 minutes)
On site facilities: A selection of souvenirs is available
Public opening: Weekdays 10.00-17.00, Sundays 14.00-17.00. Closed Good Friday, Christmas Day, Boxing Day and New Year's Day
Special Notes: The Museum also houses a considerable number of photographs depicting scenes of the GWR along with nameplates, models, posters, tickets, etc. 1991: Return to Swindon Exhibition continues this year

Locomotives

Name	No	Origin	Class	Type	Built
—	2516	GWR	2301	0-6-0	1897
Lode Star	4003	GWR	'Star'	4-6-0	1907
—	9400	GWR	9400	0-6-0PT	1947
*North Star**	—	GWR	—	2-2-2	1837
—	4	GWR	Diesel railcar	Bo-Bo	1934
—	—	A38W	Petrol railcar	2w-2PMR	c1960

*Broad gauge replica

Owners
All locomotives are part of the National Railway Museum Collection

Below:
NER 0-6-0 No 1275 was one of a number of exhibits at the 'NRM on Tour' show at Swindon during 1990. *P. Q. Treloar*

Winchcombe Railway Museum M

Three miles from the Gloucestershire-Warwickshire railway, the diverse collection includes signalling equipment, lineside fixtures, horse-drawn road vehicles, tickets, lamps, etc. Indoor and outdoor displays set in ½ acre of traditional Victorian Cotswold Garden. Visitors are encouraged to touch exhibits
Location: 23 Gloucester Street, Winchcombe, Gloucestershire

OS reference: SP 023283
Operating society/organisation: Winchcombe Railway Museum Association, 23 Gloucester Street, Winchcombe, Gloucestershire
Telephone: Winchcombe (0242) 602257/620641
Car park: On street at entrance
Access by public transport: Bus service from Cheltenham operated by Castleways Ltd

On site facilities: Relic, refreshments and souvenir shop
Public opening: Daily Easter to October 13.30-18.00. Weekends and holidays only November to March 13.30-dusk
Facilities for disabled: Access to all parts except toilets
Special notes: Many visitor-operated exhibits

South West

Avon Valley Railway SC

Formerly the Bristol Suburban Railway, services re-commenced over the Bitton-School Road Halt section in August 1983
Headquarters: Bitton Railway Co Ltd, Bitton Station, Bristol, Avon
Main station: Bitton
OS reference: ST 670705
Car park: Bitton
Access by public transport: Badgerline service No 332 (Bristol-Bath), No 558 (Bristol-North Common) or 632 (Bristol-Bath)
Catering facilities: Refreshment room is able to supply hot snacks. Restaurant coach is available for private party catering
On site facilities: Toilets, snack bar
Length of line: 1½ miles
Public opening: Please contact for details, available on pre-recorded message on Bitton (0272) 327296
Special events: Santa Specials — 23/24, 30 November, 1, 7/8, 14/15, 21-24, 26 December
Facilities for disabled: Coach converted to disabled vehicle
Membership details: Mrs V. Baldwin, 18 Gays Road, Hanham, Bristol

Locomotives

Name	*No*	*Origin*	*Class*	*Type*	*Built*
Sir Frederick Pile	34058	SR	BB	4-6-2	1947
—	44123	LMS	4F	0-6-0	1925
—	45379	LMS	5MT	4-6-0	1937
—	47324	LMS	3F	0-6-0T	1926
—	48173	LMS	8F	2-8-0	1943

Locomotive notes: All locomotives undergoing restoration.

Industrial locomotives

Name	*No*	*Builder*	*Type*	*Built*
Edwin Hulse	2	Avonside (1798)	0-6-0ST	1918
Littleton No 5	—	M/Wardle (2018)	0-6-0ST	1922
Fonmon	—	Peckett (1936)	0-6-0ST	1924
—	—	RSH (7151)	0-6-0T	1944
—	—	R/Hornsby (235519)	4wDM	1945
—	—	Baguley/Drewry (2153)	0-4-0	1941
—	2	Bagnall (2842)	0-4-0ST	1946
General Lord Robertson	610	Sentinel (10143)	0-8-0DH	1961
Kingswood	—	Barclay (446)	0-4-0DM	1959
—	10128	Sentinel (10128)	0-4-0DM	1963

Locomotive notes: R/Hornsby (235519) and Bagnall (2842) undergoing restoration.

Stock
9 ex-BR Mk1 coaches (2 stored off-site)
1 ex-BR Mk 1 Restaurant Coach
1 ex-BR Mk 1 sleeper
1 ex-GER coach chassis
1 ex-LMS 'Royal Scot' corridor 3rd
1 ex-LMS brake composite corridor
3 open wagons
3 box vans
3 refrigerated box vans
1 ex-GWR Fruit 'C'
1 ex-GWR ballast wagon
1 ex-GWR hand crane and match truck
1 ex-GCR box van
1 ex-BR Lowmac
1 ex-BR ballast wagon
1 ex-LMS 12-ton box van
1 ex-LMS 5½-ton handcrane and runner

Owners
44123 and 47324 (the London Midland Society)

Bicton Woodland Railway

A passenger-carrying line of 18in gauge with stock mainly from the Woolwich Arsenal Railway and of World War 1 vintage
Location: Bicton Park, near Sidmouth
OS reference: SY 074862
Operating society/organisation: Bicton Woodland Railway, Bicton Gardens, East Budleigh, Budleigh Salterton, Devon
Telephone: Colaton Raleigh (0395) 68465
Car park: On site
Access by public transport: Buses pass half-hourly from Exeter, Exmouth, Sidmouth in season
On site facilities: Refreshments, shop, one museum, toilets, 18in gauge railway
Length of line: 3,250yd
Public opening: Please contact for details

Locomotives

Name	*No*	*Builder*	*Type*	*Built*
Woolwich	1	Avonside (1748)	0-4-0T	1916
Bicton	2	R/Hornsby (213839)	4wDM	1942
Carnegie	3	Hunslet (4524)	0-4-4-0DM	1954
Clinton	4	H/Hunslet (2290)	0-4-0	1941
*Budley**	—	R/Hornsby (235624)	4wDM	1945

*Static exhibit

Stock
4 open bogie coaches
5 closed bogie coaches

Facilities for disabled: Toilets, wheelchairs available. Special carriage for wheelchairs

Bodmin Steam Railway

The Bodmin Steam Railway typifies the bygone branch railways of Cornwall. The terminus, close to Bodmin town centre, has an interesting collection of small standard gauge locomotives and rolling stock; and the operating line winds down to a junction with British Rail at Bodmin Parkway. From the track there are glimpses of Bodmin Moor, and scenic views across the beautiful valley of the River Fowey
Location: Bodmin General station, on B3268
Operating society/organisation: Bodmin & Wenford Railway, Bodmin General Station, Bodmin, Cornwall PL31 1AQ
Telephone: All enquiries (0208) 73666
Car park: Bodmin General only, no parking permitted at BR's Bodmin Parkway station
Access by public transport: Interchange with BR at Bodmin Parkway (arrivals by BR train only). Local bus services to Bodmin
Refreshment facilities: Light refreshments only

Locomotives

Name	*No*	*Origin*	*Class*	*Type*	*Built*
—	3802	GWR	2884	2-8-0	1938
—	5552	GWR	4575	2-6-2T	1928
Western Lady	D1048	BR	52	C-C	1962
—	W79976	BR (AC Cars)	—	4wDM	1958
—	D3452	BR	10	0-6-0DE	1957
—	D3559	BR	08	0-6-0DE	1958

Industrial locomotives

Name	*No*	*Builder*	*Type*	*Built*
—	19	Bagnall (2962)	0-4-0ST	1950
Alfred	—	Bagnall (3058)	0-4-0ST	1953
—	—	Bagnall (3121)	0-4-0F	1957
Peter	—	Fowler (22928)	0-4-0DM	1940
Progress	—	Fowler (4000001)	0-4-0DM	1945
Swiftsure	—	Hunslet (2857)	0-6-0ST	1943
—	—	Hunslet (3133)	0-4-0DM	1949
Progress	—	Peckett (1611)	0-4-0ST	1923
Corrall	—	R/Hornsby (304470)	4wDM	1946
Lec	—	R/Hornsby (443642)	4wDM	1960
Ugly	62	RSH (7673)	0-6-0ST	1950

Stock
8 BR Mk 1 coaches
3 GWR coaches
1 GWR Siphon G
Various freight wagons

Owners
D1048 (the Western Lady Ltd)

On site facilities: Souvenir shop, limited display of historic artefacts, toilets
Length of line: 3½ miles General-Parkway (track to Boscarne is *in situ* but not useable)
Passenger trains: 29-31 March; 1 April; 5-6, 26-31 May; 2, 5, 9, 12, 16-21, 23-28, 30 June; 1-5, 7-12, 14-19, 21-26, 28-31 July; daily in August; 1-13, 15-20, 22-27, 29-30 September; 6, 13, 20, 27 October; 7/8, 14/15, 21/22, 28/29 December; 1 January 1992
Facilities for disabled: Yes
Membership details: C/o above address

Bristol Industrial Museum SC

The Museum houses machinery and vehicles associated with Bristol's industrial past, from horse-drawn vehicles to aircraft
Location: Princes Wharf, Bristol
OS reference: ST 585722
Operating society/organisation: Bristol Industrial Museum, Princes Wharf, Bristol, Avon BS1 4RN
Telephone: Bristol (0272) 251470
Car parks: Available nearby
Access by public transport: Buses to centre of town
On site facilities: Shop
Length of line: ½-mile
Public opening: Saturday-Wednesday 10.00-13.00; 14.00-17.00

Industrial locomotives

Name	*No*	*Builder*	*Type*	*Built*
Portbury	S3	Avonside (1764)	0-6-0ST	1917
Henbury	S9	Peckett (1940)	0-6-0ST	1937
—	3	F/Walker (242)	0-6-0ST	1874
—	—	R/Hornsby (418792)	0-4-0DM	1958

Facilities for disabled: Reasonable access
Special notes: Operation of railway on advertised weekends only
Membership details: Officer in charge D. Martin, Bristol Harbour Railway c/o above address

Buckfast Steam Railway

A typical West Country branch line meandering up the steep Dart Valley to Buckfastleigh which is home to the railway's workshops, a butterfly and otter farm and several other attractions
Headquarters: Buckfast Steam Railway, Buckfastleigh station, Buckfastleigh, Devon
Telephone: Buckfastleigh (0364) 42338
Main station: Buckfastleigh
Other public stations: Staverton Bridge
OS reference: Buckfastleigh SX 747663, Staverton Road SX 785638
Car park: Buckfastleigh
Access by public transport: Bus, X 38/9 Exeter-Plymouth; 188 Newton Abbot-Buckfastleigh
Refreshment facilities: Buckfastleigh
Souvenir shop: Buckfastleigh
Museum: Buckfastleigh
Depot: Buckfastleigh
Length of line: 7 miles
Passenger trains: Buckfastleigh-Littlehempston Riverside alongside the River Dart

Locomotives

Name	*No*	*Origin*	*Class*	*Type*	*Built*
Bulliver	1420	GWR	1400	0-4-2T	1933
Ashburton	1450	GWR	1400	0-4-2T	1935
—	1369	GWR	1366	0-6-0PT	1934
Dartington	1638	GWR	1600	0-6-0PT	1951
—	3803*	GWR	2884	2-8-0	1939
—	6435	GWR	6400	0-6-0PT	1937
Thornbury Castle	7027	GWR	'Castle'	4-6-0	1949
Ardent	D2192	BR	03	0-6-0DM	1961
Tiny†	—	SDR	—	0-4-0VBT	1868

*Undergoing restoration at the Birmingham Railway Museum, Tyseley
†Broad gauge (7ft 0¼in)

Industrial locomotives

Name	*No*	*Builder*	*Type*	*Built*
Ashley	1	Peckett (2031)	0-4-0ST	1942
Lady Angela	1690	Peckett (1690)	0-4-0ST	1926
—	—	Bagnall (2766)	0-6-0ST	1944
Barbara	—	Hunslet (2890)	0-6-0ST	1943
Glendower	—	Hunslet (3810)	0-6-0ST	1954
—	—	Fowler (421014)	0-4-0DM	1958

Period of public operation: Telephone above for details
Facilities for disabled: Limited
Membership details: Mr A. Bissett, Dart Valley Railway Association, 98 Union Road, Shirley, Solihull, West Midlands B90 3DG

Stock
DMU vehicles, Class 127 Nos 51592 and 51604
7 ex-BR Mk 1 coaches
9 ex-GWR coaches
4 ex-GWR auto trailers
Stock is sometimes transferred to and from the Paignton & Dartmouth Steam Railway

Owners
0298 (30587) (on loan from the National Railway Museum)
Tiny, 7ft 0¼in gauge, part of the National Collection
3803 (the Dumbleton Hall Society)
Glendower, *Barbara* and Bagnall 2766 (the Glendower Group)
7027 (the Birmingham Railway Museum, being restored by Dumbleton Hall Preservation Society)

Below:
0-6-0ST *Barbara* wheels a demonstration freight through Buckfastleigh station on 29 September 1990.
Melvyn Hopwood

East Somerset Railway (The Strawberry Line)

Set up by the artist, David Shepherd, 'the man who loves giants', the railway line is home to *Black Prince* and *The Green Knight* housed in their 'traditional' shed. As one might expect, Cranmore station is well laid out and aesthetically pleasing. A new art gallery is situated at Cranmore station where his paintings and prints can be bought
Headquarters: East Somerset Railway, (Cranmore Railway Station) Shepton Mallet, Somerset
OS reference: ST 664429
Telephone: Cranmore (074 988) 417

Locomotives

Name	*No*	*Origin*	*Class*	*Type*	*Built*
—	6634	GWR	5600	0-6-2T	1928
—	32110	LBSCR	E1	0-6-0T	1877
—	47493	LMS	3F	0-6-0T	1927
The Green Knight	75029	BR	4MT	4-6-0	1954
Black Prince	92203	BR	9F	2-10-0	1959
—*	930	RR	10	4-8-0	1896

Industrial locomotives

Name	*No*	*Builder*	*Type*	*Built*
Lord Fisher	1398	Barclay (1398)	0-4-0ST	1915
Lady Nan	1719	Barclay (1719)	0-4-0ST	1920
—	4101	Dubs (4101)	0-4-0CT	1901

Main station: Cranmore
Car park: Cranmore — free
Refreshment facilities: New restaurant situated in car park offering lunches, snacks, teas, etc. Group catering by arrangement. Picnic areas at Cranmore, Merryfield Lane stations and depot
Souvenir shop: Cranmore
On site facilities: Museum, Victorian style engine shed and workshops, children's play area, wildlife information centre
Depot: Cranmore West
Length of line: 2 miles
Passenger trains: Cranmore West to Mendip Vale. Stations at Cranmore West, Merryfield Lane and Mendip Vale, unlimited train travel
Period of public operation: Depot open daily 27 March-30 October, with trains running each Sunday, Public Holidays; Wednesdays, Thursdays and Saturdays in peak season and on certain other days. Closed January and February (except 1 January). Depot open weekends only November, December and March 10.00-16.00. Last admission 30min before closing time. Each day ticket allows unlimited travel on all timetabled trains
Special events: 20 June — Jazz Night in the Sheds; 26 August — Cranmore Village Fayre; 5/6 October — Enthusiasts' Weekend; 30 November/1, 7/8, 14/15, 21/22 December — Santa Specials
Facilities for disabled: Limited, advance notice required
Special Notes: Santa special steam trains on Saturday and Sundays in December
Membership details: Please apply to above address, SAE for brochure

Stock
6 ex-BR Mk 1 coaches
1 ex-LMS coach
1 ex-LNER (design) sleeping car
25 assorted wagons, mostly LMS and SR
Rhodesian Railways sleeping car 1808*
Note: *3ft 6in gauge

Owner
6634 (the 6634 Locomotive Co)

Launceston Steam Railway

The railway runs through the beautiful Kensey Valley on a track gauge of 1ft 11½in, following the trackbed of the old North Cornwall line. The locomotives formerly worked on the Dinorwic and Penrhyn railways in North Wales. Launceston station contains a museum of vintage cars and motorcycles and there is also a collection of stationary steam engines which are demonstrated at work. There are catering and gift shop facilities and an exhibition of model railways, together with a very well stocked model shop. At the far end of the line there are pleasant walks and a picnic area. The covered rolling stock ensures an enjoyable visit whatever the weather
Location: Newport Industrial Estate, Launceston, Cornwall
OS reference: SX 328850
Operating Society/organisation: The Spice Settlement Trust Co Ltd, trading as the Launceston Steam Railway, Newport, Launceston
Telephone: (0566) 775665
Stations: Launceston-Hunts Crossing-New Mills. Note: the line now terminates ½-mile from the hamlet of New Mills and refreshments are available at the temporary terminus
Car park: Newport Industrial Estate, Launceston
Length of line: 2 miles, but 2½ miles when New Mills extension opens
Gauge: 1ft 11½in
Access by public transport: BR Gunnislake 13 miles, Plymouth or Bodmin 25 miles
On site facilities: Cafe and restaurant, transport museum, workshop tours, gift and bookshop, and model railway display with model shop, all situated at Launceston
Period of public operation: Easter holiday, then Tuesdays and Sundays until Whitsun. Daily Whitsun until end of September. Tuesdays and Sundays in October. Santa Specials every Saturday and Sunday in December, also Christmas Eve and Boxing Day
Public opening: Trains run from 11.00-17.00. Departures every 40min and more frequently if required. Unlimited riding on date of issue of ticket
Journey time: Return 35min, no single tickets available
Facilities for disabled: Easy access to all areas except bookshop and motorcycle museum. No toilet facilities for disabled
Special events: Double-headed trains on Wednesdays in July and August (whenever possible)

Industrial locomotives

Name	*No*	*Builder*	*Type*	*Built*
Lilian	—	Hunslet (317)	0-4-0ST	1883
Velinheli	—	Hunslet (409)	0-4-0ST	1886
Covertcoat	—	Hunslet (679)	0-4-0ST	1898
Sybil	—	Bagnall (1760)	0-4-0ST	1906
Dorothea	—	Hunslet (763)	0-4-0ST	1901
—	—	M/Rail (5646)	4wDM	1933
—	—	M/Rail (9546)	4wDM	1950

Locomotive notes: The three Hunslet locomotives (317/409/679) are expected to be in use during 1991.

Stock
1 electric inspection trolley
Bogie and 4-wheel coaches

Paignton & Dartmouth Steam Railway

A line that succeeds in imparting the feeling of running from A to B, linking, as it does, the Dartmouth ferry at Kingswear to Paignton and BR. A trip in the observation car is well worth the supplement for the views of the sea, the climb through the tunnel, and the descent through the wooded hillside to Brunel's terminus at Kingswear on the banks of the River Dart
Headquarters: Paignton Queen's Park station, Paignton, Devon
Telephone: Paignton (0803) 555872
Main station: Paignton Queen's Park
Other public stations: Goodrington, Churston, Kingswear (for Dartmouth)
OS reference: SX 889606
Car parks: Paignton municipal car park, Goodrington, Dartmouth (ferry to Kingswear)
Access by public transport: Adjacent to both BR station and Devon General bus station
Refreshment facilities: Kingswear
Depot: Churston
Length of line: 7 miles
Passenger trains: Paignton-Kingswear, views of Torbay and Dart estuary, 495yd tunnel
Period of public operation: Easter to October

Locomotives

Name	*No*	*Origin*	*Class*	*Type*	*Built*
—	4555	GWR	4500	2-6-2T	1924
—	4588	GWR	4575	2-6-2T	1927
Dumbleton Hall	4920	GWR	Hall	4-6-0	1929
Goliath	5239	GWR	5205	2-8-0T	1924
Lydham Manor	7827	GWR	7800	4-6-0	1950
Volunteer	D3014	BR	08	0-6-0DE	1954
Mercury	D7535	BR	25	Bo-Bo	1965

Stock
10 ex-BR Mk 1 coaches
1 ER Suburban coach (ex-King's Cross)
1 Pullman observation coach
2 DMU trailer vehicles, Class 116 Nos 59003/59004
1 auto-coach
Stock is sometimes transferred to and from the Buckfast Steam Railway

Owners
4929 (the Dumbleton Hall Society)

Facilities for disabled: Limited
Special events: Santa Specials — please see press for details
Membership details: Mrs P. Mackley, Renewals Officer, 68 Osney Crescent, Paignton TQ4 5EZ

Plym Valley Railway M

A scheme dedicated to the restoration of services over the former GWR Marsh Mills-Plym Bridge line, a distance of 1½ miles
Location: 5 miles from centre of Plymouth, Devon, north of A38. From Marsh Mills roundabout, take B3416 to Plympton, follow signs
OS reference: SX 517564
Operating society/organisation: Plym Valley Railway Co Ltd, Marsh Mills Station, Coypool Road, Marsh Mills, Plymouth, Devon PL7 4NL
Access by public transport: Buses from Plymouth, Nos 20, 20A, 21, 22A, 51. Stop close to site
On site facilities: Shop and refreshments at Marsh Mills, Coypool (weekends only)
Public opening: Site open most days from 10.00

Locomotives

Name	*No*	*Origin*	*Class*	*Type*	*Built*
Wadebridge	34007	SR	WC	4-6-2	1945
—	75079	BR	4MT	4-6-0	1956
—	D3002	BR	08	0-6-0DE	1953

Industrial locomotives

Name	*No*	*Builder*	*Type*	*Built*
—	3	H/Leslie (3597)	0-4-0ST	1926
—	—	T/Hill (125V)	4wDH	1963

Special events: Please see press for details
Special notes: Visitors are advised that, at the moment, the railway and its locomotives are still under restoration. Prospective visitors are advised to take the advice of their guides. Steam and diesel operating on various weekends
Membership details: Marsh Mills Station, Coypool Road, Marsh Mills, Plymouth PL7 4NL

Seaton & District Electric Tramway

A passenger-carrying 2ft 9in gauge electric tramway, operating on the trackbed of the former LSWR Seaton branch
Location: Seaton, Devon
OS reference: SY 252904
Operating society/organisation: Seaton & District Electric Tramway Co, Riverside Depot, Harbour Road, Seaton, Devon EX12 2NQ
Telephone: Seaton (0297) 21702/ 20375
Access by public transport: Located near to centre of Seaton

Locomotives

Name	*No*	*Builder*	*Type*	*Built*
—	—	R/Hornsby (435398)	4wDM	1959

On site facilities: Shops Seaton and Colyton, tea bar at Colyton, adventure playground
Length of line: 3 miles, 2ft 9in gauge
Period of public operation: Daily Good Friday-end September, weekends in October, also limited winter service

Special notes: Services operated by open top deck bogie tramcars and enclosed single-deck winter car
Disclaimer: The company reserves the right to cancel the tram service without giving prior notice, although this would apply in extreme circumstances only

Swanage Railway — The Purbeck Line

Overlooked by the historic ruins of Corfe Castle, this railway is slowly extending towards Wareham and a connection to the BR network
Location: Swanage station
Operating society/organisation: Swanage Railway Co Ltd, Station House, Railway Station, Swanage, Dorset
Telephone: Swanage 425800. Purbeck Line 24hr Talking Timetable — Swanage 424276
Main station: Swanage
Other public stations: Herston Halt, Harman's Cross
OS reference: SZ 026789
Car park: Swanage
Access by public transport: Wilts & Dorset buses from Bournemouth, Poole and Wareham stop outside station
On site facilities: Souvenir shop, buffet car on certain trains, hot and cold snacks. Shop open every weekend throughout the year, and on all operating days
Length of line: 3 miles, Swanage-Herston Halt-Harman's Cross
Public opening: Weekends throughout the year. Daily from June to September (inclusive). Thursdays in April, May and October. DEMU early and late trains from July to September
Special events: 31 March/1 April — Easter Egg Specials; 6/7 April — Friends of Thomas the Tank Engine Weekend; 20/21 April — Teddy Bear Specials; 5 May — Vintage Vehicle Rally; 18/19 May — Spring Steam

Locomotives

Name	No	Origin	Class	Type	Built
—	5619	GWR	5600	0-6-2T	1924
—	6695	GWR	5600	0-6-2T	1928
—	30053	LSWR	M7	0-4-4T	1905
—	120	LSWR	T9	4-4-0	1899
257 Squadron	34072	SR	BB	4-6-2	1948
Holland America Line	35022	SR	MN	4-6-2	1948
Port Line	35027	SR	MN	4-6-2	1948
—	45160	LMS/WD	8F	2-8-0	1941
—	80078	BR	4MT	2-6-4T	1954
—	80104	BR	4MT	2-6-4T	1955
—	1708	MR	IF	0-6-0T	1880
—	—	JZ	USA	0-6-0T	1950s
—	D3591	BR	08	0-6-0DE	1958
—	D7594 (25244)	BR	25	Bo-Bo	1964

Locomotive notes: 5619 is on loan until May 1991. 30053 is due to re-enter service in spring 1991 following restoration at Swindon. 34072 will be away periodically from May on short-term loans. 35027 is on loan to the Bluebell Railway. 45160 is undergoing restoration at Poole. 80104 is at Swindon Heritage Centre undergoing restoration

Industrial locomotives

Name	*No*	*Builder*	*Type*	*Built*
Cunarder	47160	Hunslet (1690)	0-6-0T	1931
Linda	21	H/Leslie (3931)	0-6-0ST	1938
Nunlow	—	H/Clarke (1704)	0-6-0T	1938
May	2	Fowler (4210132)	0-4-0DM	1957
Beryl	—	Planet (2054)	4wPM	1937
—	2150	Peckett (2150)	0-6-0T	1954

Stock
Hastings line DEMU (3 coaches)
3 ex-LSWR coach bodies
4 ex-SR vans
9 ex-SR coaches
18 ex-BR Mk 1 coaches

Gala; 1/2 June — Friends of Thomas the Tank Engine Visit the Seaside; 22/23 June — Victorian Weekend; 18/19 September — 1940s Weekend; 28/29 September — Autumn Steam Gala; 26-27 October — Friends of Thomas the Tank Engine Weekend; Weekends in December — Santa Specials 26-31 December — Mince Pie Specials; 1 January 1992 — New Year's Day Specials
Facilities for disabled: Access to station, shop and toilets and special coach on train
Feature article reference: RW December 1989, p733 (photo feature)

1 ex-LNWR CCT
1 ex-LMS coach
18 various types of wagons
1 ex-BR Sleeping coach
1 ex-SR 15ton diesel-electric crane
1 Wickham trolley (WD9024)
1 ex-London Transport Plasser & Theurer ballast tamper
1 ex-BR Corridor 2nd, converted to disabled persons coach

Owners
120 (the National Railway Museum)
5619 (the Telford Steam Railway)
6695 (the Great Western Railway Preservation Group)
80078 and 35022 (the Southern Steam Trust)
Cunarder (owned by 1708 Locomotive Preservation Trust Ltd)
34072, 35027 and 80104 (the *Port Line* Project)
Nunlow (the Bahamas Locomotive Society)

Tiverton Museum M

The Museum, dominated by No 1442, affectionately known as the 'Tivvy Bumper', houses a large collection of railway relics
Location: Tiverton, Devon
OS reference: SS 955124
Operating society/organisation: Tiverton Museum Society, St Andrew Street, Tiverton, Devon EX16 6PH
Telephone: Tiverton (0884) 256295
Car park: Adjoining

Locomotives

Name	*No*	*Origin*	*Class*	*Type*	*Built*
—	1442	GWR	1400	0-4-2T	1935

Access by public transport: Rail to Tiverton Parkway, then by bus, or bus from Exeter
On site facilities: Museum, shop and toilets
Public opening: Daily 10.30-16.30; except Sundays and from 24 December to 31 January
Special notes: Limited facilities for disabled

West Somerset Railway

Running for 20 miles, this is Britain's longest preserved railway and evokes all the atmosphere of a country railway from a more leisured age. The line is host to several societies and groups, and several stations have their own museum, such as the Somerset and Dorset Trust; at Washford. There are some idyllic country stations in the Quantock Hills and Dunster station served for many years as the model for Hornby Dublo's branch line station
Headquarters: West Somerset Railway, The Railway Station, Minehead, Somerset
Telephone: Minehead (0643) 4996
Main station: Minehead
Other public stations: Dunster, Blue Anchor, Washford, Watchet, Williton, Doniford Beach Halt, Stogumber, Crowcombe, Bishop's Lydeard

Locomotives

Name	*No*	*Origin*	*Class*	*Type*	*Built*
—	53808	S&DJR	7F	2-8-0	1925
—	3205	GWR	2251	0-6-0	1946
—	3850	GWR	2884	2-8-0	1942
—	4160	GWR	5101	2-6-2T	1948
—	4561	GWR	4500	2-6-2T	1924
—	5542*	GWR	4575	2-6-2T	1928
—	6412	GWR	6400	0-6-0PT	1934
Dinmore Manor	7820†	GWR	'Manor'	4-6-0	1950
Western Yeoman	D1035	BR	52	C-C	1962
—	D2271	BR	04	0-6-0DM	1952
—	D2994	BR	07	0-6-0DE	1962
—	D7017	BR	35	B-B	1962
—	D9500	BR	14	0-6-0DH	1964
—	D9526	BR	14	0-6-0DH	1964
—	D9551	BR	14	0-6-0DH	1965

*Undergoing restoration at Taunton
†Undergoing restoration at Birmingham Railway Museum, Tyseley

OS reference: Minehead SS 975463, Williton ST 085416, Bishops Lydeard ST 164290
Car parks: Minehead, Williton, Bishops Lydeard. Some parking at all stations except Washford and Doniford Beach
Access by public transport: Nearest BR station, Taunton
Refreshment facilities: Minehead, Bishop's Lydeard (limited opening), Williton (limited opening). Wine and Dine trains. Buffet car on most steam trains
Souvenir shops: Minehead, Bishop's Lydeard
Museum: Somerset & Dorset Railway Museum Trust, Washford. GWR Museum at Blue Anchor
Depots: Bishop's Lydeard, Williton, Washford, Minehead
Length of line: 20 miles
Passenger trains: Steam and diesel trains to Bishops Lydeard
Period of public operation: SAE for details
Special events: Gala Week, 1-8 September
Facilities for disabled: Parking space level with entrance. No steps to shop or booking office, level access to toilets (disabled toilets at Minehead should be commissioned during 1990). Catering facilities can be reached without difficulties. Groups can be catered for. Advanced booking advised to guarantee facilities. No special facilities on train but assistance available
Feature article references: RW September 1989, p541 (photo feature); RW July 1990, p376 (Diesel & Electric Group)

Industrial locomotives

Name	*No*	*Builder*	*Type*	*Built*
Portbury	S3	Avonside (1764)	0-6-0ST	1917
—	—	Bagnall (2473)	0-4-0ST	1932
*Jennifer**	20	H/Clarke (1731)	0-6-0T	1942
Isabel	—	H/Leslie (3437)	0-6-0ST	1919
*Kilmersdon**	—	Peckett (1788)	0-4-0ST	1929
—	24	Ruston (210479)	4wDM	1941
—	—	Ruston (183062)	4wDM	1937
—	57	Sentinel (10214)	0-6-0DM	1964

*May not be on site for all of 1991 season

Stock
20 ex-BR Mk 1 coaches
1 ex-BR Mk 2 coach
2 ex-BR Restaurant cars
7 ex-GWR camping coaches
1 ex-BR Mk 1 Sleeper
1 ex-GWR Sleeping Coach
1 ex-SR 'Ironclad' coach
2 Park Royal 2-car diesel multiple-units (50415, 50414, 56168, 56169)
2 Gloucester 2-car diesel multiple-units (50341, 51118, 56097, 56099)
1 Cravens 2-car diesel multiple-unit (51485, 56121)
1 ex-GWR Hawksworth coach
1 ex-GWR 5-ton hand crane
More than 40 freight vehicles

Owners
53808, S3, *Isabel, Kilmersdon* and Bagnall 2473 (the Somerset & Dorset Museum Trust)
D7017 and D9526 (the Diesel and Electric Group)
5542 (the 5542 Fund)
3205 (the 2251 Class Fund)
3850 and 7820 (the 3850 Preservation Society)
4160 (4160 Ltd)
D1035 (Foster Yeoman Ltd — actually D1010)

Below:
Stogumber is awakened from its slumbers as S&DJR No 53808 rolls in with a train bound for Bishop's Lydeard on 16 July 1990. *Wayne Finch*

Wales

Bala Lake Railway (Rheilffordd Llyn Tegid)

Although narrow gauge, this railway is built over the old standard gauge branch to Bala, which is the line's north-eastern terminus. The railway's headquarters are to be found in the fine old station building at Llanuwchllyn at the other end of the line. Do not be deterred by the fact that the railway runs down the opposite shore of the lake to the main road — it is well worth the detour
Headquarters: Rheilffordd Llyn Tegid (Bala Lake Railway) Llanuwchllyn, Bala, Gwynedd LL23 7DD
Telephone: Llanuwchllyn (067 84) 666
Main station: Llanuwchllyn
Other public stations: Llangower, Bala
OS reference: Llanuwchllyn SH 880300, Bala SH 929350
Car parks: Llanuwchllyn, Llangower and Bala town centre
Access by public transport: Bus Gwynedd service No 94 to both Bala and Llanuwchllyn (from Wrexham or Barmouth)
Road access: Off the A494 Bala-Dolgellau road
Refreshment facilities: Llanuwchllyn. Large picnic site with toilet facilities by lake at Llangower
Souvenir shop: Llanuwchllyn
Museum: Large collection of antique railway equipment
Depot: Llanuwchllyn
Length of line: 4½ miles, 1ft 11⅝in gauge
Passenger trains: Llanuwchllyn-Bala. Journey takes 25min in each direction
Period of public operation: 23 March-29 September
Facilities for disabled: Facilities available on most trains
Special notes: Small parties (10/12) may just turn up, but a day's notice for large parties would be helpful
Membership details: P. Briddon, 140 Earl Marshal Road, Sheffield S4 8LB

Above:
***Holy War* trundles along the edge of Bala Lake on its journey from Bala to Llanuwchllyn on 7 May 1990.** *Doug Birmingham*

Industrial locomotives

Name	*No*	*Builder*	*Type*	*Built*
Holy War	—	Hunslet (779)	0-4-0ST	1902
Maid Marian	—	Hunslet (822)	0-4-0ST	1903
Meirionnydd	—	Severn Lamb (7322)	Bo-Bo	1973
Chilmark	—	R/Hornsby (194771)	4wDM	1939
Indian Runner	—	R/Hornsby (200744)	4wDM	1940
—	—	Lister (34025)	4wDM	1949
—	—	Hudson (38384)	4wDM	1930
—	—	Motorail (5821)	4wDM	1934
—	—	R/Hornsby (209430)	4wDM	1942
—	—	R/Hornsby (189972)	4wDM	1938
—	—	Hibberd (FH2544)	4wDM	1941
—	—	Hunslet (1974)	4wDM	1939
Cernyw	—	R/Hornsby (200748)	4wDM	1940

Locomotive notes: *Holy War* and *Maid Marian* are in regular use, remainder are on static display

Brecon Mountain Railway

A narrow gauge passenger carrying railway close to Merthyr Tydfil built on part of the trackbed of the former Brecon & Merthyr Railway. Gradually being extended northward, the railway has some interesting narrow gauge steam locomotives imported from East and West Germany and South Africa
Headquarters: Brecon Mountain Railway, Pant Station, Dowlais, Merthyr Tydfil, Mid Glamorgan CF48 2UP
Telephone: Merthyr Tydfil (0685) 4854
Main station: Pant
Car park: Pant station
OS reference: SO 063120
Access by public transport: Omnibus to Pant Cemetery — ½ hourly frequency from Merthyr bus station. BR rail service to Merthyr from Cardiff Central
Depot: Pant
Length of line: 2 miles, 1ft 11¾in gauge
Period of public operation: Easter weekend; May weekend, daily 25 May-mid-September
Refreshment facilities: Licensed restaurant at Pant, snackbar at Pontsticill

Locomotives

Name	*No*	*Builder*	*Type*	*Built*
—	—	Baldwin (61269)	4-6-2	1930
Sybil	—	Hunslet (827)	0-4-0ST	1903
San Justo	4	H/Clarke (639)	0-4-2ST	1903
Santa Ana	—	H/Clarke (640)	0-4-2ST	1903
Graf Schwerin-Löwitz	99.3353	Arn Jung (1261)	0-6-2WT	1908
—	—	O&K (12722)	0-4-0WT	1936
Pendyffryn	—	de Winton	0-4-0VBT	1894
Redstone	—	Redstone	0-4-0VBT	1905
Rhydychen	—	Simplex (11177)	4wDM	1961
Garret	—	M/Rail (7902)	4wDM	1939
—	77	Hanomag (10629)	2-6-2+2-6-2	1928
—	—	Brecon MR (001)	0-6-0DH	1987
—	—	Brecon MR (003)	0-6-0DH	1987

Stock
9 bogie wagons
2 balcony end 39-seat coaches
2 balcony end 40-seat coaches
1 20-seat Caboose
Miscellaneous rail carrying and ballast wagons
South African Railways Box Car
Diesel-hydraulic crane
Wickham petrol trolley

Special events: Santa Specials 7/8, 14/15, 21/22 December
Facilities for disabled: Toilets and ramps

Caerphilly Railway Society

SC

An unusual collection in the historic surroundings of the old Rhymney Railway locomotive works, comprising a rare selection of locomotives and rolling stock from both main line railways and South Wales industry.

The restored Brecon & Merthyr Railway signalbox, with its unique overlocking Saxby & Farmer lever frame, is worth seeing
Operating society: Caerphilly Railway Society Ltd
Location: Harold Wilson Industrial Estate, Van Road, Caerphilly, Mid Glamorgan
OS reference: ST 163865
Telephone: 0222 461414 (office hours)
Car parking: Ample car parking available at the depot
Public transport: Approximately 1 mile from Caerphilly rail and bus stations

Locomotives

Name	*No*	*Origin*	*Class*	*Type*	*Built*
—	41312	LMS	2MT	2-6-2T	1952
—	28	TVR	01	0-6-2T	1897
—	D2178	BR	03	0-6-0DM	1962

Industrial locomotives

Name	*No*	*Builder*	*Type*	*Built*
Victory	—	Barclay (2201)	0-4-0ST	1945
Forester	—	Barclay (1260)	0-4-0ST	1911
Desmond	—	Avonside (1498)	0-4-0ST	1906
Haulwen	—	V/Foundry (5272)	0-6-0ST	1945
—	—	Hunslet	0-6-0DM	1962
Deighton	—	YEC (2731)	0-4-0DE	1959

Locomotive notes: Locomotives operating from the 1990 season are expected to be TVR 0-6-2T No 28 and *Victory*. Locomotives currently under repair are D2178, 41312 and *Deighton*. All others currently static/in store pending restoration.

On site facilities: Light snacks, refreshments and souvenirs available on steam days
Facilities for the disabled: No steps are involved in gaining access to the depot. Facilities are limited, but help readily given
Length of line: ¾-mile
Public opening: Static display every Sunday 14.00-17.00
Special events: For 1991 steam days please see railway press
Membership: Mr N. Radley, 3 Warwick Place, Cardiff, South Glamorgan

Stock
GWR 6-ton hand crane and flat runner wagon
GWR 14-ton ballast wagon 60501
3 wooden 7-plank mineral wagons
2 steel-framed oil tank wagons
BR Mk 1 BSK E34460 (internally converted to buffet facility)
GWR 20-ton Toad goods brake van 35267
SR 4-wheel PMV 1168 (latterly mess and tool van)
GWR/BR Fruit D W92040
LMS 12-ton boxvan 304986
GWR Mink D 28804 (latterly mess and tool van)
Swansea Harbour Trust 4-wheel saloon carriage
TVR coach (body only) No 153

Owners
28 (the National Railway Museum, on loan from the Museum of Wales)
D2178 and Hunslet (on loan from National Smokeless Fuels)

Conwy Valley Railway Museum M

Conveniently situated alongside British Rail's Betws-y-Coed station, the Museum presents some well-displayed distractions to pass the time including model train layouts to delight both adult and child
Location: Adjacent to Betws-y-Coed station
OS reference: SH 796565
Operating society/organisation: Conwy Valley Railway Museum, The Old Goods Yard, Betws-y-Coed, Gwynedd
Telephone: (06902) 568
Car park: On site
Access by public transport: Betws-y-Coed BR station
On site facilities: Refreshments in buffet car. Bookshop and model/gift shop in museum foyer, operating train layouts, miniature railway (1¼-miles, 7¼in gauge) steam-hauled. Picnic area

Locomotives

Name	*No*	*Builder*	*Type*	*Built*
Sgt Murphy	—	K/Stuart (3117) (2ft gauge)	0-6-0T	1918
Britannia	70000	TMA Engineering (1ft 3in gauge)	4-6-2	1988

Stock
1 GWR fitter's van
1 LMS 6-wheel van
1 LNER CCT van
1 BR Mk 1 coach
2 SR luggage vans
1 Pullman coach

Public opening: Daily Easter-end of October, 10.30-17.30
Facilities for disabled: Access to museum and toilets

Corris Railway Museum M

This rather remote Museum merits a detour if only because of the unusual treatment of and detailed research into local history with some fascinating film and photographs that can be seen in the old stone building that houses the Museum
Location: Off A487, opposite Braichgoch Hotel, five miles north of Machynlleth
OS reference: SH 755078

Locomotives

Name	*No*	*Builder*	*Type*	*Built*
Alan Meaden	5	M/rail (22258)	4wDM	1965
—	6	R/Hornsby (51849)	4wDM	1966

Locomotive notes: 5 operational and on works trains, 6 currently under repair

Stock
Small number of representative and works wagons, one manriding carriage, with a second currently having a replica body added

Operating society: The Corris Railway Society, Corris Station Yard, Gwynedd (postal address: Corris, Machynlleth, Powys SY20 9SH)
Car park: Adjacent
Access by public transport: Bus Gwynedd services 34 & 35 (Machynlleth-Aberllefeni and Machynlleth-Dolgellau-Blaenau Ffestiniog respectively) and 'Trawscambria' (stop at Braichgoch Hotel)
Catering facilities: None, but light teas available nearby in village
On site facilities: Souvenir shop, toilets and children's playground
Length of line: ¾-mile, 2ft 3in gauge construction line, being upgraded to passenger carrying standard
Public opening: Please contact for details
Special events: Please contact for details
Facilities for disabled: Limited
Membership details: A. H. Lawson, 165 Gynsill Lane, Anstey, Leicester LE7 7AN

Fairbourne & Barmouth Railway

Since 1983 this railway has now been regauged from 15in to 12¼in but has been transformed by the introduction of new locomotives and rolling stock, tunnels through the sand dunes, miniaturised stations and signalboxes, new workshops, a créperie overlooking the Barmouth estuary and a butterfly jungle alongside Fairbourne station
Headquarters: North Wales Narrow Gauge Railway Co Ltd, Fairbourne & Barmouth Leisure Park, Beach Road, Fairbourne, Gwynedd, LL38 2EX
Telephone: 0341-250362 or 250083
Main station: Gorsaf Newydd, Fairbourne
Other public stations: Gorsafawddacháidraigd-danheddogleddolôn-penrhynareudraethceredigion, Porth Penrhyn
OS reference: SH 616128
Car parks: Gorsaf Newydd, Gorsa-fawddach
Access by public transport: Fairbourne BR station. Bus Gwynedd service
Refreshment facilities: Celtic Pavillion at Porth Penrhyn. Porth Penrhyn restaurant specialises in Welsh Lamb and Breton Pancakes, with 'Sundowner' evening service for dinner and entertainment (main season) (phone for details)

Locomotives

Name	*No*	*Builder*	*Type*	*Built*
Beddgelert	—	Curwen	0-6-4T	1979
Yeo	—	Curwen	2-6-2T	1978
Sherpa	—	Milner	0-4-0STT	1978
Russell	—	FLW	2-6-2T	1985
Lilian Walter	—	FLW	A1-1AD	1985
Gwril	—	FLW	4wBE	1987
Sandy River	24	FLW	2-6-2	1989

FLW — Fairbourne Locomotive Works

Stock
12¼in gauge — 28 coaches (1st, 2nd class)
16 freight

Note: At the time of compilation, the line was 'for sale' with no purchaser announced. Please see press for details or contact before visit

Souvenir shop: Gorsaf Newydd and Penrhyn Pavillion
Depot: Fairbourne
Length of line: 2¾ miles, 12¼in gauge
Passenger trains: A 2¾-mile journey connecting with ferry at Porth Penrhyn to Barmouth. 20min single journey. Through tickets to Barmouth at reduced price
Period of public operation: Daily Good Friday-1 November (closed Saturdays except at peak season)
Facilities for disabled: Full disabled facilities with Ro-Ro coach
Membership details: Friends of Fairbourne — contact Hon Sec A. de Frayssinet at above address
Special notes: During inclement weather the service may be restricted or cancelled. Extra trains and special parties by arrangement with the manager. Fairbourne Butterfly Jungle at Gorsaf Newydd

Ffestiniog Railway

In many ways, evocative of the early Swiss mountain railways as it climbs high above Porthmadog with some breathtaking views, the railway still operates an interesting variety of locomotives including some unusual Victorian survivors. Passengers have replaced slate as the principal traffic over this former quarry line

Locomotives

Name	*No*	*Builder*	*Type*	*Built*
Princess	1	G/England (199/200)	0-4-0STT	1863
Prince	2	G/England	0-4-0STT	1863
Welsh Pony	—	G/England (234)	0-4-0STT	1867
Earl of Merioneth	—	FR	0-4-4-0T	1979
Merddin Emrys	10	FR	0-4-4-0T	1879
Moelwyn	—	Baldwin (49604)	2-4-0DM	1918

Headquarters: Ffestiniog Railway, Harbour Station, Porthmadog, Gwynedd, LL49 9NF
Telephone: Porthmadog (0766) 512340 or 831654
Main stations: Porthmadog Harbour, Blaenau Ffestiniog
Other public stations: Minffordd, Penrhyn, Tan-y-Bwlch, Tanygrisiau
OS reference: SH 571384
Car parks: Porthmadog, Tan-y-Bwlch, Tanygrisiau, Blaenau Ffestiniog
Access by public transport: Minffordd and Blaenau Ffestiniog BR stations. Porthmadog, Minffordd, Tanygrisiau and Blaenau Ffestiniog served by local buses
Refreshment facilities: Licensed restaurant at Porthmadog, buffet at Tan-y-Bwlch (summer only), also on most trains
Souvenir shops: Porthmadog, Tan-y-Bwlch, Blaenau Ffestiniog
Museum: Porthmadog
Depot: Boston Lodge
Length of line: 13½ miles, 1ft 11½in gauge
Passenger trains: Porthmadog-Blaenau Ffestiniog
Period of public operation: Weekends March. Daily April-October, and Christmas
Special events: Santa Specials — please enquire for details
Facilities for disabled: Porthmadog and Blaenau Ffestiniog easily accessible for wheelchairs. Facilities on trains for disabled in wheelchairs by prior arrangement
Special notes: Reduced return rates available for journeys beginning on diesel services, shown in timetable
Feature article references: RW October 1989, p618 (FR showing the way); RW April 1990, p209 (Upgrading the Image)

Blanche	—	Hunslet (589)	2-4-0STT	1893
Linda	—	Hunslet (590)	2-4-0STT	1893
*Britomart**	—	Hunslet (707)	0-4-0ST	1899
Mountaineer	—	Alco (57156)	2-6-2T	1917
Livingston Thompson†	3	FR	0-4-4-0T	1886
—†	K1	B/Peacock (5292)	0-4-0+0-4-0	1909
Harlech Castle	—	Baguley-Drewry (3767)	0-6-0-DH	1983
Upnor Castle	—	Hibberd (3687)	4wDM	1954
Conway Castle	—	Hibberd (3831)	4wDM	1958
Moel Hebog	—	Hunslet (4113)	0-4-0DM	1955
Mary Ann	—	M/Rail (596)	4wDM	1917

7 other 4wDM units
*Privately owned
†On loan to National Railway Museum

Stock
27 bogie coaches
4 4-wheel coaches
2 brake vans, plus numerous service vehicles

Below:
Activity at Tan-y-Bwlch as double Fairlie *Merddin Emrys* arrives with a Porthmadog-bound service. *Mountaineer* heads the up train to Blaenau Ffestiniog on 24 October 1990. *P. Q. Treloar*

Great Orme Tramway

A cable tramway to the summit of the Great Orme is operated as two sections involving a change half-way. Opened throughout in July 1903, it involves gradients as steep as 1 in 10.3
Location: Great Orme Tramway, Victoria station, Church Walks, Llandudno
OS reference: SH 7781
Operating society/organisation: Aberconwy Borough Council, Direct Services Organisation, Maesdu, Llandudno LL30 1HF
Telephone: Llandudno (0492) 870870
Car park: Approximately 100yd from Lower Terminal. Adjacent to Summit Terminal
Access by public transport: Good
On site facilities: Shop
Period of public operation: Easter to end of September (daily)
Special notes: Cable tramway to summit of Great Orme, approximately 1-mile rising to 650ft (3ft 6in gauge)
Stock: 4 tramcars each seating 48, built 1902/3

Gwili Railway (Rheilffordd Gwili)

South Wales' first standard gauge passenger-carrying railway extended services back to Llwyfan Cerrig during 1987
Headquarters: Gwili Railway Co Ltd, Bronwydd Arms station, Bronwydd Arms, Carmarthen, Dyfed
Telephone: Carmarthen (0267) 230666
OS reference:
Bronwydd Arms SN 417239
Llwyfan Cerrig SN 405258,
Conwil SN 386263
Main station: Bronwydd Arms
Other public stations: Llwyfan Cerrig, also Conwil under restoration
Car park: Bronwydd Arms
Access by public transport: Carmarthen BR station, then Crosville bus service No 400 (not Sundays).
Refreshment facilities: Bronwydd Arms, Llwyfan Cerrig (picnic site). Bar on train
Souvenir shop: Bronwydd Arms
Depot: Llwyfan Cerrig, stock also kept at Bronwydd Arms, Conwil and 'Ironrails', Carmarthen
Length of line: 1¾ miles
Passenger trains: Bronwydd Arms-Llwyfan Cerrig, regular hourly service
Period of public operation: Please contact for details

Industrial locomotives

Name	*No*	*Builder*	*Type*	*Built*
—	1	H/Clarke (1885)	0-6-0ST	1955
Little Lady	1903	Peckett (1903)	0-4-0ST	1936
Idris	—	R/Hornsby (207103)	4wDM	1941
Trecatty	—	R/Hornsby (421702)	0-6-0DM	1959
Olwen	7058	RSH (7058)	0-4-0ST	1942
—	71516	RSH (7170)	0-6-0ST	1944
Nellie	D2875	YEC(2779)	0-4-0DE	1960
Rosyth No 1	1	A/Barclay (1385)	0-4-0ST	1914*
Swansea Vale No 1	1	Sentinel (9622)	4wVBTG	1958*
Swansea Jack	—	R/Hornsby (393302)	4wDM	1955*
Dylan Thomas	—	N/British (27654)	0-4-0DH	1956*
—	—	R/Hornsby (3307)	4wDM	1938

Stock
1 ex-BR diesel unit trailer car
4 ex-BR Mk 1 coaches
2 ex-BR suburban coaches
1 ex-BR griddle car
1 ex-TVR coach (built 1891)
Coles diesel rail crane
1 ex-GWR Mink van
1 ex-GWR Fruit D
1 ex-GWR Monster
1 ex-GWR Crocodile
1 ex-GWR Loriot D
1 tank wagon
2 GWR Toad brake vans
1 GWR Tube C
1 GWR Tunnel Toad
14 open wagons

Owners
*The Railway Club of Wales

Public opening: 11.00-17.15 on all operating days
Special events: See press for details

Llanberis Lake Railway (Rheilffordd Lyn Padarn)

A narrow gauge passenger carrying railway starting at the historic Dinorwic Quarry workshops (now part of the National Museum of Wales) and running along the shores of the Llanberis lake using the trackbed of the former slate railway line to Port Dinorwic. Excellent views of Snowdonia and good picnic spots along the line
Headquarters: Llanberis Lake Railway, Gilfach Ddu, Llanberis, Gwynedd LL55 4TY
Telephone: Llanberis (0286) 870549
Main station: Llanberis (Padarn station/Gilfachddu)
Other public stations: Cei Llydan
OS reference: SH 586603
Car park: Llanberis (Padarn station)

Industrial locomotives

Name	No	Builder	Type	Built
Elidir	1	Hunslet (493)	0-4-0ST	1889
Thomas Bach/Wild Aster	2	Hunslet (849)	0-4-0ST	1904
Dolbadarn	3	Hunslet (1430)	0-4-0ST	1922
Helen Kathryn	5	Henschel (28035)	0-4-0T	1948
—	7	R/Hornsby (441427)	4wDM	1961
Twll Coed	8	R/Hornsby (268878)	4wDM	1956
Dolgarrog	9	M/Rail (22154)	4wDM	1962
—	—	R/Hornsby (425796)	4wDM	1958
Garrett	11	R/Hornsby (198286)	4wDM	1939
Braich	10	R/Hornsby (203031)	4wDM	1942
—	18	M/Rail (7927)	4wDM	1941
Llanelli	19	R/Hornsby (451901)	4wDM	1961
*Una**	—	Hunslet (873)	0-4-0ST	1905

*Not part of the railway's motive power stock. Housed at the adjacent slate museum and can sometimes be seen working demonstration freight trains

Stock
13 bogie coaches
20 wagons

Refreshment facilities: Padarn station
Souvenir shop: Padarn Station
Length of line: 2 miles, 1ft 11½in gauge
Passenger trains: Llanberis-Penllyn-Llanberis
Journey time: 40min round trip
Period of public operation: Mondays to Thursdays in March and October. Monday to Friday in April and May. Sundays to Fridays, June through September, Saturdays July and August
Facilities for disabled: Level approaches throughout shop, cafe and to train. Special toilet facilities provided. All disabled visitors welcomed

Llangollen Railway SC

The line, which is presently 3½ miles long, is the only preserved standard gauge line in North Wales. Situated in the Dee Valley, it follows the course of the River Dee for much of its route, and affords good views of the surrounding countryside between Llangollen and Deeside Loop. It is the eventual aim to reach Corwen, some 9 miles from Llangollen where a terminus will be re-established
Location: Llangollen station, A542 from Ruabon, A5 from Shrewsbury, A5 from Betws y Coed
Operating society/organisation: Llangollen Railway Society Ltd, Llangollen station, Llangollen, Clwyd
Telephone: Answerphone (24hr): Llangollen (0978) 860951. Other enquiries: (0978) 860979 (office hours only)
Main station: Llangollen
Other stations: Berwyn, Deeside Loop
OS reference: SJ 214422
Car park: Llangollen (Market Street)
Access by public transport: Nearest station: Ruabon (2hr service), then Crosville bus service No D1 to Llangollen or Nos D93/D94
Refreshment facilities: Llangollen and Berwyn
Souvenir shop: Llangollen
Length of line: 3½ miles
Passenger trains: Llangollen-Deeside Loop
Period of public operation: Saturdays and Sundays Easter-July; daily July-mid-September; Saturdays and Sundays mid-September-mid-October. Special event days during winter as advertised. Please see timetable for details of services to Deeside Loop

Locomotives

Name	*No*	*Origin*	*Class*	*Type*	*Built*
—	2859	GWR	2800	2-8-0	1918
Kinlet Hall	4936†	GWR	'Hall'	4-6-0	1929
—	5199	GWR	5101	2-6-2T	1934
—	5538	GWR	4575	2-6-2T	1928
Cogan Hall	5952	GWR	'Hall'	4-6-0	1935
—	6430†	GWR	6400'	4-6-0PT	1937
—	7754	GWR	5700	0-6-0PT	1930
Ditcheat Manor	7821	GWR	'Manor'	4-6-0	1950
Foxcote Manor	7822	GWR	'Manor'	4-6-0	1950
Odney Manor	7828	GWR	'Manor'	4-6-0	1950
—	03162	BR	03	0-6-0DM	1960
—	D3265 (08195)	BR	08	0-6-0DE	1956
—	D7629	BR	25	Bo-Bo	1965
—	25313	BR	25	Bo-Bo	1966
—	D9500*	BR	14	0-6-0DH	1964
—	D9502	BR	14	0-6-0DH	1964

Also ex-BR Class 127 No M51618 and Class 105 No 54456 DMUs
*Undergoing restoration at West Somerset Railway
†Not on site

Right:
The 14.00 Llangollen-Deeside Loop service on 27 May 1990 was headed by No 7822 *Foxcote Manor* which is seen here departing Berwyn loop.
Doug Birmingham

Special events: Please refer to timetable or event leaflet
Facilities for disabled: Special passenger coach for wheelchairs, also shop and refreshment rooms at Llangollen. Toilet available at Berwyn station. Advance notice required for special coach
Membership details: Mr J. Short, 'Bryn Aber', Llangollen Road, Trevor, Nr Llangollen, Clwyd
Feature article reference: RW October 1989, p605 (photo feature — Great Western Steam at . . .)

Industrial locomotives

Name	*No*	*Builder*	*Type*	*Built*
Darfield No 1	—	Hunslet (3783)	0-6-0ST	1953
Eliseg	—	Fowler (22753)	0-4-0DM	1939
—	—	Fowler (400007)	0-4-0DM	1947
Richboro	—	H/Clarke (1243)	0-6-0T	1917
Burtonwood Brewer	—	Kitson (5459)	0-6-0ST	1932
—	14	H/Clarke (D1012)	0-4-0DM	1956
—	—	North British (27734)	0-4-0DH	1958
—	1	YEC/BTH	0-4-0D	c1950
—	—	YEC	0-6-0DE	—
—	—	R/Hornsby (416213)	0-4-0DE	1957

Stock
17 Mk 1 coaches
2 Mk 1 sleepers
4 GWR coaches
4 wagons
1 Bolster wagon
1 LNWR tool van
2 GWR brake vans
1 GWR Mink D wagon
1 SR 'BY' parcels van
2 tank wagons
1 LNER parcels van
2 BR Fruit vans
1 ex-LNWR brake van
1 Matisa track tamper
1 BR ballast wagon
1 Coles diesel/electric 5-ton crane
1 BR generator van
1 ex-Inspection Machine
1 GWR Siphon G coach
1 LMS box van
1 GWR Mink A van
1 BR CCT
1 Matisa Track Recording Machine
1 BR Bridge/Viaduct Inspection Unit
1 steam crane

Owners
5199, 5952, 7821 and 7828 (the GW Steam Locomotive Group)
7822 (the Foxcote Manor Society)
7754 (the National Museum of Wales)
Richboro (the National Coal Board)
Burtonwood Brewer (the Burtonwood Brewery)
03162 (the Wirral Borough Council)

Narrow Gauge Railway Centre M

A very extensive collection of narrow gauge railway equipment from the British Isles mainly housed in a purpose-built exhibition hall at the former Oakley Slate Quarries. Close to other tourist attractions in Blaenau Ffestiniog, principally the Llechwedd Slate Caverns and the Ffestiniog Railway
Location: Off the A470 Blaenau Ffestiniog road ½-mile north of the town, turn at the locomotive
OS reference: SH 693470
Operating society/organisation: Gloddfa Ganol Slate Mine, Blaenau Ffestiniog, Gwynedd, North Wales LL41 3NB
Telephone: (0766 830) 664
Car park: On site
Access by public transport: Blaenau Ffestiniog BR/FR station, bus connection to mine
On site facilities: Licensed restaurant, snack bar and grill. Toilets, children's playground and playroom, slate works, craftshops, mining museum, preserved quarrymen's cottages. Extensive underground workings, Land Rover tours and rail ride
Public opening: Easter-September, Monday-Friday 10.00-17.30; Sundays mid-July-August
Length of line: ½-mile to mine entrance
Facilities for disabled: Toilets, shops, museums, restaurant and section of mine suitable. No advance notice required. Special party rates available

Penrhyn Castle Industrial Railway Museum M

A collection of historic industrial steam locomotives, both standard and narrow gauge, displayed in Penrhyn Castle, a well known National Trust property in the area regularly open to visitors
Location: Llandegai, near Bangor. One mile east of Bangor on the A5
OS reference: SH 603720
Operating society/organisation: National Trust, Penrhyn Castle, Industrial Railway Museum, Llandegai, near Bangor, Gwynedd
Telephone: Bangor (0248) 353084
Car park: Within castle grounds
Access by public transport: Nearest BR station, Bangor
On site facilities: The castle is open to the public, and contains a gift shop. Light refreshments are available
Public opening: Daily 28 March-3 November
Facilities for disabled: Access to castle and museum
Special notes: For those interested in stately homes the castle is well worth a visit. The entrance fee covers both the castle and the railway exhibits housed in the castle courtyard

Industrial locomotives

Name	*No*	*Builder*	*Type*	*Built*
Kettering Furnaces No 3	—	B/Hawthorn (859)	0-4-0ST	1885*
Watkin	—	de Winton	0-4-0VBT	1893*
Fire Queen	—	Horlock	0-4-0	1848†
Hawarden	—	H/Clarke (526)	0-4-0ST	1899
Vesta	—	H/Clarke (1223)	0-6-0T	1916
Charles	—	Hunslet (283)	0-4-0ST	1882‡
Hugh Napier	—	Hunslet (855)	0-4-0ST	1904‡
—	1	Neilson (1561)	0-4-0WT	1870
Haydock	—	Stephenson (2309)	0-6-0T	1879
—	—	R/Hornsby (327904)	0-4-0DM	1951

*3ft gauge
†4ft gauge
‡1ft 10¾in gauge

Stock

10 narrow gauge rolling stock exhibits from the Padarn/Penrhyn system. The small relics section includes a comprehensive display of railway signs and model locomotives.

Pontypool & Blaenavon Railway SC

The historic Blaenavon site, complete with its railway installations and locomotives can easily be included in a visit to Big Pit Mining Museum
Location: Near Big Pit, Blaenavon, Gwent
OS reference: SO 237093
Operating society/organisation: Pontypool & Blaenavon Railway Co (1983) Ltd, Council Offices, Lion Street, Blaenavon, Gwent
Telephone: (0495) 772726 (evenings only)
Car park: Adjacent to railway terminus
Access by public transport: Blaenavon-Brynmawr buses
On site facilities: Light refreshments and souvenir shop
Public opening: Sundays 5 May-1 September, plus Bank Holiday Mondays, 12.00-17.00. Easter train service 29-31 March, 1 April. Santa Specials 14/15, 21/22 December. Mince Pie Specials 28/29 December
Special notes: The railway incorporates the former mineral/ LNWR passenger lines running through Big Pit. Both north and southbound extensions are being considered. Service currently operates between Furnace Sidings platform and Whistle Inn platform
Membership details: c/o above address

Locomotives

Name	*No*	*Origin*	*Class*	*Type*	*Built*
—	2874	GWR	2800	2-8-0	1918
—	3855	GWR	2884	2-8-0	1942
—	4253	GWR	4200	2-8-0T	1917
—	5668	GWR	5600	0-6-2T	1926
Bickmarsh Hall	5967	GWR	'Hall'	4-6-0	1937

Industrial locomotives

Name	*No*	*Builder*	*Type*	*Built*
Brookfield	—	Bagnall (2613)	0-6-0PT	1940
Nora	5	Barclay (1680)	0-4-0ST	1920
Menelaus	—	Peckett (1889)	0-6-0ST	1935
—	8	RSH (7139)	0-6-0ST	1944
—	1	Fowler (22497)	0-6-0DM	1938
—	—	Drewry (2252)	0-6-0DH	1948
—	10083	R/Royce (10083)	0-4-0DH	1961

Stock

1 brake van
5 BR Mk 1 coaches
Several vans, china clay, coke and tank wagons

Snowdon Mountain Railway

The only public rack and pinion railway in the British Isles, this bustling line climbs up the slopes of Snowdon, often through the clouds, to the hotel at the top. The trip should not be missed
Headquarters: Snowdon Mountain Railway, Llanberis, Gwynedd LL55-4TY
Telephone: Llanberis (0286) 870223
Main station: Llanberis
Other public stations: Summit
OS reference: SH 582597
Car park: Llanberis
Access by public transport: Bangor BR station then by Crosville bus to Caernarfon and there change to local bus to Llanberis
Refreshment facilities: Llanberis, Summit
Souvenir shops: Llanberis, Summit
Depot: Llanberis
Length of line: 4¾ miles, 80cm gauge
Passenger trains: Llanberis-Summit. Journey time 60 minutes. Departures from Llanberis at 30 minute intervals during peak periods

Locomotives

Name	*No*	*Builder*	*Type*	*Built*
Enid	2	SLM (924)	0-4-2T	1895
Yr Wyddfa	3	SLM (925)	0-4-2T	1895
Snowdon	4	SLM (988)	0-4-2T	1896
Moel Siabod	5	SLM (989)	0-4-2T	1896
Padarn	6	SLM (2838)	0-4-2T	1922
Ralph	7	SLM (2869)	0-4-2T	1923
Eryri	8	SLM (2870)	0-4-2T	1923
Ninian	9	Hunslet (9249)	0-4-0D	1986
Yeti	10	Hunslet (9250)	0-4-0D	1986

All steam locomotives were built by Swiss Locomotive Works, Winterthur

Stock
8 closed bogie coaches
1 bogie works car
1 4-wheel open wagon

Period of public operation: Daily 15 March-1 November
Special notes: All trains are subject to weather and traffic restrictions, especially before mid-May and during October. Parties welcome by prior arrangement
Feature article reference: RW March 1989, p144

Swansea Maritime & Industrial Museum M

This Museum houses a number of relics from Swansea's industrial and maritime past and includes some railway exhibits, some of which are operated on the former dock sidings adjacent to the Museum. The new Tramshed Annexe houses a restored Swansea City double-deck tram (Brush Electrical Engineering 1923/24 model), and a replica of the early Mumbles Railway carriage, in addition to the sole surviving example of the Mumbles Railway rolling stock, the front section of the tramcar, manufactured by the Brush Electrical Engineering Co in 1928.

Artifacts from the Mumbles Railway are also displayed within a graphic presentation, as are examples of Brunels 'GWR' broad gauge rail track.
Location: On the south side of the town between the shopping centre and the sea in the newly created Maritime Quarter
OS reference: SS 659927
Operating society/organisation: City of Swansea Museum Services. Maritime & Industrial Museum, Museum Square, Maritime Quarter, Swansea SA1 1SN
Telephone: Swansea (0792) 650351
Car park: Public car parks close by
Access by public transport: Reached on foot from shopping centre, central bus depot or by car
On site facilities: No refreshments in the Museum but several cafes close by. Museum shop selling souvenirs and produce of the Woollen Mill which operates in the Museum throughout the year. Education Service available on request to Education Officer
Public opening: 10.30-17.30 seven days a week, closed 25/26/27 December and New Year's Day
Facilities for disabled: Available
Membership details: Museum's Friends Organisation c/o above address

Industrial locomotives

Name	*No*	*Builder*	*Type*	*Built*
Sir Charles	—	A/Barclay (1473)	0-4-0F	1919
—	—	Peckett (1426)	0-6-0ST	1916

Talyllyn Railway

The very first railway in the country to be rescued and operated by enthusiasts, the line climbs from Tywyn through the wooded Welsh hills past Dolgoch Falls to Nant Gwernol. The trains are hauled by a variety of veteran tank engines all immaculately maintained by the railway's own workshops at Tywyn Pendre

Headquarters: Talyllyn Railway Co, Wharf station, Tywyn, Gwynedd LL36 9EY
Telephone: Tywyn (0654) 710472
Main station: Tywyn Wharf
Other public stations: Tywyn Pendre, Rhydyronen, Brynglas, Dolgoch Falls, Abergynolwyn, Nant Gwernol
OS reference: SH 586005 (Tywyn Wharf)
Car parks: Tywyn Wharf, Dolgoch, Abergynolwyn
Access by public transport: Tywyn BR station. Bus Gwynedd services to Tywyn
Refreshment facilities: Tywyn Wharf, Abergynolwyn hot and cold snacks available

Locomotives

Name	*No*	*Builder*	*Type*	*Built*
Talyllyn	1	F/Jennings (42)	0-4-2ST	1865
Dolgoch	2	F/Jennings (63)	0-4-0WT	1866
Sir Haydn	3	Hughes (323)	0-4-2ST	1878
Edward Thomas	4	K/Stuart (4047)	0-4-2ST	1921
Midlander	5	R/Hornsby (200792)	4wDM	1940
Douglas	6	Barclay (1431)	0-4-0WT	1918
Tom Rolt	7	Barclay (2263)	0-4-2T	1949
Merseysider	8	R/Hornsby (476108)	4wDH	1964
Alf	9	Hunslet (4136)	0-4-0DM	1950

Locomotive notes: In service — 1, 2, 3 and 4; limited use — 6; No 7 is expected to enter regular service in early summer 1991

Stock
13 4-wheel coaches/vans
10 bogie coaches
45 wagons

Narrow Gauge Museum, Tywyn

Name	*No*	*Builder*	*Type*	*Built*
Dot	—	B/Peacock (2817)	0-4-0ST	1887
*Pet**	—	LNWR	0-4-0ST	1865
Rough Pup	—	Hunslet (541)	0-4-0ST	1891
—	2	K/Stuart (721)	0-4-0WT	1902
Jubilee 1897	—	M/Wardle (1382)	0-4-0ST	1897
George Henry	—	de Winton	0-4-0T	1877
—	13	Spence	0-4-0T	1895
Nutty	—	Sentinel (7701)	0-4-0VB	1929

*On loan from National Railway Museum (not currently on site)

Stock
Various wagons and miscellaneous equipment

Below:
The empty stock of a Talyllyn Railway special is seen at Tywyn Pendre headed by No 1 *Talyllyn*. *C. V. Awdry*

Souvenir shops: Tywyn Wharf, Abergynolwyn
Museum: Tywyn Wharf
Depot: Tywyn Pendre
Length of line: 7½ miles, 2ft 3in gauge
Passenger trains: Tywyn-Nant Gwernol
Period of public operation: Daily late March to end of October, Christmas and New Year.
Journey times: Tywyn-Nant Gwernol — single 51min, return 1hr 10/20min
Special events: 25/26 May — Vintage Transport Rally, Tywyn; 17 August — Unusual 'Race the Train' event; 25 August — Land Rover Rally, Tywyn; 21 December — Santa Specials
Facilities for disabled: No problem for casual visitors, advanced notice preferred for groups. Access to shop and cafeteria possible at Tywyn and Abergynolwyn. Disabled toilet facilities at Abergynolwyn. Access possible to lower floor of museum. Limited capacity for wheelchairs on trains. New vehicle for wheel chairs now in operation on selected trains
Special notes: Special parties and private charter trains by arrangement. Children under 5 years of age free. Narrow gauge 'Wanderer' four- and eight-day tickets accepted
Membership details: Mr A. Johnston, 9 Reynolds Way, Croydon, Surrey CR0 5JW
Feature article reference: RW January 1990, p13

Vale of Rheidol Railway

This narrow gauge railway offers a 23-mile round trip from Aberystwyth to Devils Bridge providing spectacular views which cannot be enjoyed by road. At Devils Bridge there are walks to the Mynach Falls and Devils Punch Bowl. Many artists have been inspired by the magnificence of Devils Bridge and the Rheidol Valley
Operating company: Brecon Mountain Railway, Pant station, Dowlais, Merthyr Tydfil, Mid Glamorgan CF48 2UP
Telephone: Merthyr Tydfil (0685) 4854
Main station: Aberystwyth (adjacent to BR station)
Other public stations: Devil's Bridge, Rhiwfron, Rheidol Falls, Aberffrwd, Nantyronen, Capel Bangor, Glanrafon, Llanbadarn
OS reference: SN 587812
Car parks: Aberystwyth, Devil's Bridge

Locomotives

Name	*No*	*Origin*	*Ex-BR Class*	*Type*	*Built*
Owain Glyndwr	7	GWR	98	2-6-2T	1923
*Llywelyn**	8	GWR	98	2-6-2T	1923
Prince of Wales	9	GWR	98	2-6-2T	1924
—	10	Brecon MR (002)	98/1	0-6-0DH	1987

*Currently runs unnamed in Cambrian Railways livery

Stock
16 bogie coaches including one Vista coach
1 4-wheel baggage van
12 wagons for maintenance use
1 inspection trolley

Access by public transport: Aberystwyth BR station: and bus services to Aberystwyth
Refreshment facilities: Aberystwyth (not railway-owned), Devils Bridge (not railway operated)
Souvenir shop: Aberystwyth
Depot: Aberystwyth (not open to the public)
Length of line: 11¾ miles, 1ft 11½in gauge
Passenger trains: Aberystwyth-Devil's Bridge
Period of public operation: Daily 29 March-30 September
Feature article reference: RW January 1991, p14

Wales Railway Centre SC

The world's first steam locomotive ran in Wales in 1804 and the aim of the Wales Railway Centre is to show the history of railways in Wales and their impact on the country. It is situated at Bute Road station which was once the headquarters of the Taff Vale Railway; the first main line company in Wales. The station building houses the Railway Gallery of the Welsh Industrial & Maritime Museum, while outside, work has started on the restoration of 10 ex-Barry locomotives, all of which are representatives of

Locomotives

Name	*No*	*Origin*	*Class*	*Type*	*Built*
—	2861	GWR	2800	2-8-0	1918
—	4115	GWR	4101	2-6-2T	1936
—	5227	GWR	5205	2-8-0T	1924
—	5539	GWR	4575	2-6-2T	1928
—	6686	GWR	5600	0-6-2T	1928
Willington Hall	7927	BR	'Hall'	4-6-0	1950
—	44901	LMS	5MT	4-6-0	1945
—	48518	LMS	8F	2-8-0	1944
—	80150	BR	4MT	2-6-4T	1956
—	92245	BR	9F	2-10-0	1958

locomotives which worked in different parts of the Principality. It is hoped that ex-Mountain Ash Colliery 0-6-0ST *Sir Gomer* will be working by the summer of 1991
Location: Bute Road Railway station, Cardiff
Operating society: The Butetown Historic Railway Society
Car park: On site
Access by public transport: Frequent train services from Cardiff (Queen Street) station (not Sundays). Frequent bus services from the city centre
On site facilities: Shop and light refreshments at weekends

Industrial locomotives

Name	*No*	*Builder*	*Type*	*Built*
Sir Gomer	—	Pecket (1859)	0-6-0ST	1932
—	52/001	Barclay (1966)	0-4-0F	1929
—	107	North British (27932)	0-6-0DM	1959
—	D1198	English Electric (D1198)	0-6-0DH	1967

Stock
Various freight vehicles

Public opening: Daily (except Mondays), 10.00-16.30; Sunday 14.30-16.30

Welsh Highland Railway SC

The 1964 company operates services over a short section at the south-western end of the old Welsh Highland line which linked the North Wales line with the Cambrian Coast. Negotiations are in hand to extend the line northward through the more spectacular scenery that attracted passengers from far and wide in the early part of the century. In the meantime, you have the compensation of low fares

Locomotives

Name	*No*	*Builder*	*Type*	*Built*
Moel Tryfan	—	Bagnall (2875)	0-4-2T	1948
Snowdon Ranger	—	Bagnall (3050)	0-4-2T	1953
Russell	—	Hunslet (901)	2-6-2T	1906
Pedemoura	—	O&K (10808)	0-6-0WT	1924
Karen	—	Peckett (2024)	0-4-2T	1942

Below:
Quietly resting between duties, *Karen* and *Katherine* are seen at Porthmadog on 13 July 1990. *Edward Garner*

Location: Porthmadog, Gwynedd, adjacent to BR station
OS reference: SH 571393
Operating society/organisation: Welsh Highland Light Railway (1964) Ltd, Gelert's Farm Works, Madoc Street West, Porthmadog, Gwynedd LL49 9DY
Telephone: Porthmadog (0766) 513402 (weekends and operating days) or 051-327 3576 (evenings) or 051-608 2696 (day)
Car park: At terminus
Catering facilities: Station buffet — 'Russells' supplying hot meals, cold buffet, sandwiches and light refreshments
Access by public transport: Rail service to Porthmadog BR station. Festiniog Railway Harbour station. Bws Gwynedd service 1 and 3 to Porthmadog
On site facilities: Souvenir shop, information boards, taped commentary on coaches, sheds and workshops open to visitors. Afternoon tea served on 15.00, 15.45 and 16.30 trains on Wednesdays during June and September
Length of line: ¾-mile, 1ft 11½in gauge

Glaslyn	1	R/Hornsby (297030)	4wDM	1952
Kinnerley	2	R/Hornsby (354068)	4wDM	1953
—	—	R/Hornsby (191658)	4wDM	1938
—	—	R/Hornsby (237914)	4wDM	1946
Cnicht	36	M/Rail (8703)	4wDM	1941
LBLR	9	M/Rail (9547)	4wDM	1950
Katherine	—	M/Rail (60S363)	4wDM	1968
Pilkingtons No 4	—	M/Rail (60S333)	4wDM	1959
Pilkingtons No 6	—	M/Rail (11165)	4wDM	1968
Felin Foel	—	Hunslet (6285)	4wDM	1968

Locomotive notes: 1991 steam service will be worked by *Russell*, the only surviving locomotive from the original Welsh Highland Railway, *Karen* will be the stand-by engine. Diesel service will be worked by *Glaslyn* with *Kinnerley* on stand-by.

Stock
A collection of rolling stock including bogie coaches and numerous service construction vehicles including Vale of Rheidol brake van (1902)

Passenger trains: Porthmadog-Pen-y-Mount. Return journey time 30 minutes approximately, steam hauled bank holidays, weekends during high season and every day (except Mondays and Fridays) in August
Period of public operation: Please contact for details
Special events: Please contact for details
Facilities for disabled: Disabled passengers can be accommodated without prior notice — except in parties
Membership details: Membership Secretary, c/o above address
Feature article reference: RW May 1990, p306 (*Russell*)

Welshpool & Llanfair Light Railway

There is a decidedly foreign atmosphere to the trains over this line. The steam locomotive collection embraces examples from three continents, and the coaches are turn-of-the-century balcony saloons from Austria or 1960s bogies from Africa. The line follows a steeply graded route (maximum 1 in 24) through very attractive rolling countryside, and is rather a gem in an area too often missed by the traveller heading for further shores
Headquarters: Welshpool & Llanfair Light Railway Preservation Co Ltd, Llanfair Caereinion station, Powys
Telephone: Llanfair Caereinion (0938) 810441
Main station: Llanfair Caereinion
Other public stations: Castle Caereinion, Sylfaen, Welshpool (Raven Square)
OS reference: SJ 107069
Car park: Llanfair Caereinion, Welshpool (both free)
Access by public transport: BR station at Welshpool, one mile from Raven Square. Crosville buses from Shrewsbury, Oswestry and Newtown

Locomotives

Name	*No*	*Builder*	*Type*	*Built*
The Earl	1	B/Peacock (3496)	0-6-0T	1902
The Countess	2	B/Peacock (3497)	0-6-0T	1902
Monarch	6	Bagnall (3024)	0-4-4-0T	1953
Chattenden	7	Drewry (2263)	0-6-0DM	1949
Dougal	8	Barclay (2207)	0-4-0T	1946
Sir Drefaldwyn	10	Franco-Belge (2855)	0-8-0T	1944
Ferret	11	Hunslet (2251)	0-4-0DM	1940
Joan	12	K/Stuart (4404)	0-6-2T	1927
SLR 85	14	Hunslet (3815)	2-6-2T	1954
Orion	15	Tubize (2369)	2-6-2T	1948

Locomotive notes: Locomotives expected in service 1991 — *Countess, Joan, Sir Drefaldwyn, Dougal. The Earl* will be on display at the Didcot Railway Centre. The remainder can be seen at Llanfair station, 15 is displayed with access to the footplate.

Stock
1 Wickham trolley
6 W&LLR wagons
8 ex-Admiralty wagons
2 ex-Bowater wagons
5 ex-Zillertalbahn coaches
4 ex-Sierra Leone coaches

Above:
The Bodmin Steam Railway opened its extension through to BR's Bodmin Parkway station in 1990 and on Sunday 2 September Nos 19 and D3559 were operating the day's service. We see them here leaving Bodmin General station with an afternoon service.
Alan C. Butcher

Below:
Middleton Moor Road station on 17 June 1990. *Primrose No 2*, on loan from the Embsay Steam Railway is seen in company with Middleton's former LNER Class Y1 Sentinel. During 1990 the Middleton Railway celebrated 30 years of public opening.
Robin Stewart-Smith

Refreshment facilities: Light refreshments at Llanfair Caereinion
Souvenir shops: Llanfair Caereinion
Depot: Llanfair Caereinion
Length of line: 8 miles, 2ft 6in gauge
Passenger trains: Welshpool-Llanfair Caereinion
Period of public operation: Weekends and Bank Holidays Easter to early October. Daily mid-July-mid-September
Special events: Santa trains on three weekends before Christmas; 20/21 April — Children's Weekend; 22/23 June — Friends of Thomas the Tank; 31 August/1 September — Garden Railways Exhibition; 5/6 October — Postman's Express
Facilities for disabled: Specially adapted coach for wheelchairs now available. Easy access to shop, but no special toilet arrangements
Membership details: John Parkinson, 124 London Road, Long Sutton, Spalding, Lincolnshire PE12 9EE
Feature article reference: RW February 1990, p110 (Raven Square station)

Right:
Making an impressive exit with the midday train on 15 September 1990 from Raven Square station at Welshpool is No 12 *Joan*. Built in 1927 the Kerr Stewart 0-6-2T saw service in the sugar plantations of the West Indies before arriving at the Welshpool & Llanfair Light Railway in 1971. *Edward Garner*

West Midlands

Birmingham Museum of Science & Industry — M

Location: Newhall Street, Birmingham B3 1RZ
OS reference: SP 064874
Organisation: Birmingham City Council
Telephone: 021-236 1022
Car park: Public multi-storey nearby in Newhall Street
Access by public transport: Birmingham New Street or Snow Hill stations, then short walk following signs
Facilities: Shop, refreshments
Opening times: Monday-Saturday 09.30-17.00; Sunday 14.00-17.00
Special notes: Facilities for the disabled (toilets and lifts). Special parking facility

Locomotives

Name	*No*	*Origin*	*Class*	*Type*	*Built*
City of Birmingham	46235	LMS	8P	4-6-2	1939

Industrial locomotives

Name	*No*	*Builder*	*Type*	*Built*
Secundus	—	B/Seekings	0-6-0WT	1874*
Leonard	1	Bagnall (2087)	0-4-0ST	1919†
Lorna Doone	56	K/Stuart (4250)	0-4-0ST	1922†

*2ft 8in gauge
†2ft gauge

Birmingham Railway Museum — SC

The old motive power depot is slowly being transformed into an imaginative museum of local railway history. The climax of any visit must be the view from the elevated walkway in the engine shed over the locomotives below, particularly if one has been lit up in preparation for a tour over the main line and is gently hissing steam
Location: 670 Warwick Road (A41), Tyseley, Birmingham B11 2HL
OS reference: SP 105841
Operating society/organisation: Birmingham Railway Museum Trust
Telephone: 021-707-4696
Car park: Site
Access by public transport: West Midlands Travel bus routes Nos 37, 44 from city centre. BR service to Tyseley station (no Sunday service)
On site facilities: The Museum is on the site of a former GWR/BR steam shed and has been equipped with much specialised railway engineering

Locomotives

Name	*No*	*Origin*	*Class*	*Type*	*Built*
Albert Hall	4983	GWR	'Hall'	4-6-0	1931
Earl of Mount Edgcumbe	5043	GWR	'Castle'	4-6-0	1936
Defiant†	5080	GWR	'Castle'	4-6-0	1939
Thornbury Castle‡	7027	GWR	'Castle'	4-6-0	1949
Clun Castle	7029	GWR	'Castle'	4-6-0	1950
—	7752†	GWR	5700	0-6-0PT	1930
—	7760	GWR	5700	0-6-0PT	1930
—	9600	GWR	5700	0-6-0PT	1945
*Kolhapur**	5593	LMS	'Jubilee'	4-6-0	1934
Scots Guardsman	6115	LMS	'Royal Scot'	4-6-0	1927
—	13029	BR	08	0-6-0DE	1953
—	27059	BR	27	Bo-Bo	1962
—	40118	BR	40	1Co-Co1	1961

*On loan to the Great Central Railway
†On loan to Gloucestershire Warwickshire Railway
‡On loan to Buckfast Steam Railway

Locomotive notes: 4983 is undergoing restoration, 9600 is dismantled

Left:
On 19 May 1990, No 2 *The Countess* prepares to leave Welshpool with a special charter to Llanfair Caereinion. The Welshpool & Llanfair Light Railway travels over the undulating countryside of the River Banwy valley, a trip on the open veranda of the ex-Zillertalbahn coaches will not be forgotten in a hurry. *Michael Baker*

Below:
***Lilian*, a 1883-built Hunslet 0-4-0ST is seen leaving Launceston on 18 August 1990. This little line, along with its associated motorcycle museum, makes an interesting diversion off the A30 trunk road. The Launceston Steam Railway is home to four of these delightful engines.** *Hugh Ballantyne*

Above right:
***Montalban* is a 0-4-0T built by Orenstein & Kopple in 1913 and is usually resident on the West Lancashire Light Railway, but during September 1990 it took a trip down to Bedfordshire for the Leighton Buzzard Railway's enthusiast weekend. *Montalban* is seen at Page's Park station being prepared for the day's duties.** *Robin Stewart-Smith*

Below right:
The new 15in gauge Bure Valley Railway was opened on 10 July 1990 using two steam engines borrowed from the Romney, Hythe & Dymchurch Railway. Here we see *Samson* and *Winston Churchill* being readied to work the official opening special. *Alan C. Butcher*

No22
MONTALBAN

Wroxham

machinery. Souvenir shop, restaurant, passenger demonstration line and station, viewing gallery, schools education service
Refreshment facilities: Available in 'Chuffs' restaurant
Length of line: ⅓-mile
Public opening: Static display daily 10.00-17.00 except Christmas and New Year. Steam days most Sundays and Bank Holidays, Easter-October
Special events: Santa and Halloween (see press for details)
Special notes: Tyseley is a centre for 'Steam on BR' railtours over the BR lines to Stratford upon Avon and Didcot (via Oxford). In 1983 an Education Service for schools began operating, providing a service of visits to schools in the Midlands and party visits from schools to the museum. Membership is available to the public providing free entry to site events, magazines, members' evenings, etc.
Membership facilities: From the museum office
Feature article references: RW February 1988, p94 ('Bloomer' project — also RW April 1988, p226; RW June 1988, p360; RW August 1988, p482; RW August 1989, p501; RW November 1989, p667); RW January 1991
Note: All attractions and facilities are advertised subject to availability

Industrial locomotives

Name	*No*	*Builder*	*Type*	*Built*
Rocket	—	Peckett (1722)	0-4-0ST	1926
—	1	Peckett (2004)	0-4-0ST	1942
—	—	Baguley (800)	0-4-0PE	1920
Henry	—	H/Leslie (2491)	0-4-0ST	1901
Count Louis†	—	B/Lowke (32)	4-4-2	1924

†15in gauge — on static display

Owners
27059 (Sandwell District Council)

Stock
1 Pullman bar car*
1 Gresley buffet car
1 LNWR semi-Royal Saloon*
1 GWR VIP Saloon
1 30-ton steam crane and sundry goods vehicles
1 GWR Inspection Saloon
1 Tourist 3rd open coach
2 TPO vehicles (1 LNWR, 1908; 1 LMS 1950)
1 GWR brake 3rd corridor coach
1 BR Mk 1 BFK coach
1 LMS 50ft parcels van
1 LNWR 6-wheel guard's van
1 GWR 'Toad' guard's van
1 SR PMV
1 GWR 6-wheel Mess Tool van
Plus varied goods vehicles
*In store

Bulmer Railway Centre SC

The Centre, opened in 1968, is principally a standard gauge museum and home of two main line express locomotives *King George V* and *Princess Elizabeth*. Both are normally maintained in full working order and, as such, can be absent on main line operations elsewhere. The centre also houses a collection of other locomotives, including industrial, as well as various items of rolling stock.
Location: H. P. Bulmer Ltd (Cider Makers), Whitecross Road, Hereford. The site is ½-mile from city centre on A438 Hereford-Brecon road
OS reference: SO 505402
Operating society: The 6000 Locomotive Association in conjunction with the Princess Elizabeth Locomotive Society and the Worcester Locomotive Society, on behalf of H. P. Bulmer Ltd
Telephone: Whitchurch (Avon) (0272) 834430 for enquiries. NB: code will change to 0275 during 1991
Car park: Free parking on site

Locomotives

Name	*No*	*Origin*	*Class*	*Type*	*Built*
King George V	6000	GWR	'King'	4-6-0	1927*
—	5786	GWR	5700	0-6-0PT	1930
Princess Elizabeth	6201	LMS	7P	4-6-2	1933*
Cider Queen	2 (D2758)	Hunslet (6999)	05	0-6-0DH	1955

*Telephone to check presence on site

Industrial locomotives

Name	*No*	*Builder*	*Type*	*Built*
Pectin	—	Peckett (1579)	0-4-0ST	1921
Carnarvon	47	Kitson (5474)	0-6-0ST	1934
—	1254	H/Clarke	0-6-0	1962

Locomotive notes: During 1991, *King George V* will almost certainly be absent undergoing a major overhaul elsewhere. Similarly, No 6201 will be away on main line operations during part of the year. No 5786 and *Pectin* will carry out the official steaming requirements at the Centre.

Stock
Gresley Buffet coach
Several ex-BR Mk 1 coaches
Pullman car from ex-SR 6-PUL electric unit
Small number of freight vehicles including GWR tank wagon and BR Conflat

Access by public transport: Midland Red bus services and local private bus services. Hereford (BR) station (1 mile)
On site facilities: Souvenir/book stall. Light refreshment facilities generally available
Length of line: About ¾-mile
Public opening: Weekends and bank holidays Easter/April-September inclusive. Static display (14.00-17.00). Steam-operated brake van trips, on specified Sundays and bank holidays (11.00-17.00). Telephone for details. Closed on normal weekdays (Mon-Fri)
Special events: Steam open days Easter Sunday and Monday and last Sunday in September. Santa Specials in December (enquire for details)
Special notes: Centre for steam events and servicing depot for main line steam operations in the area
Facilities for disabled: Wheelchair access to most of the site , train rides need assistance (available), no special toilet facilities
Feature article reference: RW March 1988, p166 (locomotive 6201)
Membership details: 6000 Locomotive Association, 8 Chancel View, Abbots Mead, Hereford HR2 7XD

Owners
6201 (the Princess Elizabeth Locomotive Society)
6000 (on loan from the National Railway Museum)
5786 and 47 (the Worcester Locomotive Society)
Pectin and 1254 (the 6000 Locomotive Association)
2 (D2758) (H. P. Bulmer Ltd)

Cadeby Light Railway SC

A new museum was opened in 1990, 'The Boston Collection', encompassing the lifetime collection of Teddy Boston and his family. The narrow gauge railway running in the grounds of the old rectory has been saved by Teddy Boston's enterprising widow and a small band of dedicated supporters in the face of considerable odds. Echoes of the *Titfield Thunderbolt* and Ancient and Modern. Their endeavours deserve your support
Location: One mile south of Market Bosworth on A447
OS reference: SK 426024
Operating society/organisation: Mrs J. A. Boston, The Old Rectory, Cadeby, Nuneaton, Warwicks CU13 0AS
Telephone: Market Bosworth (0455) 290462
Car park: In local side roads
Access by public transport: Leicester City, 178 from Hinckley, Leicester Corp/Gibson Bros (Comfort Coaches) from Leicester to Market Bosworth (1¼-miles away)
On site facilities: 2ft gauge railway, also traction engines and a model railway
Refreshment facilities: Light refreshments available
Public opening: 2nd Saturday of every month, plus 1st Saturday in November and Boxing Day or by arrangement
Special events: Bonfire Night Steaming 2 November; Santa Special 14 and 26 December

Locomotives

Name	*No*	*Builder*	*Type*	*Built*
Pixie	—	Bagnall (2090)	0-4-0ST	1919
—	2	O&K (7529)	0-4-0WT	1914
—	—	Baguley (1695)	0-4-0PM	1928
New Star	—	Lister (4088)	4wPM	1931
—	20	M/Rail (8748)	4wDM	1942
—	42	M/Rail (7710)	4wDM	1939
—	87004	M/Rail (2197)	4wDM	1922
—	87009	M/Rail (4572)	4wDM	1929
—	—	M/Rail (5853)	4wDM	1934
—	—	H/Clarke (D558)	4wDM	1930
—	87008	R/Hornsby (179870)	4wDM	1936
—	87051	R/Hornsby (404967)	4wDM	1957

Standard gauge

—	—	Peckett (2012)	0-4-0ST	1942

Stock
Penrhyn Quarrymans coach
2 flat trucks
7 open trucks
2 platelayers trolleys

Special notes: Party bookings by prior arrangement

34072
TARMAC
SWINDON

Above:
Tarmac sponsored the restoration of Bulleid 'Battle of Britain' class 4-6-2 No 34072 *257 Squadron* which was returned to service during the late summer of 1990. Following a handful of private light engine runs around Kent, its home territory, No 34072 was briefly transferred to the Bluebell Railway to enable it to haul passenger trains. It later moved to its home line — the Swanage Railway. No 34072 climbs towards Three Arches bridge on 23 September 1990. *A. R. Butcher*

Below:
The 1985 edition of *Railways Restored* showed photographs of three locomotives being repatriated from Greece. The last to enter service was this American-built 'S160' class 2-8-0 which re-entered service on the Mid-Hants in May 1990. Restored as No 701 *Franklin D. Roosevelt* it is seen here on 8 December 1990 after arrival at 'Santaland' following its trip over the 'Alps'. *John H. Bird*

Cambrian Railways Society SC

Location: Oswestry station yard, Oswald Road, Oswestry, Shropshire
OS reference: SJ 294297
Operating society/organisation: Cambrian Railways Society Ltd, C. W. Mottram, 'Delamere', Old Chirk Road, Gobowen, Oswestry, Shropshire SY11 3LH
Telephone: (0691) 661648
Car park: In Society's depot
Length of line: 400yd, opening 1991 (subject to obtaining Light Railway Order)
Public opening: Wednesday 10.00-17.00 (market day) and Sunday 11.00-17.00 throughout the year

Industrial locomotives

Name	*No*	*Builder*	*Type*	*Built*
—	1872	B/Peacock (1827)	0-4-0ST	1879
—	1	H/Clarke (D843)	0-4-0DM	1954
Adam	1	Peckett (1430)	0-4-0ST	1916
—	3	Hunslet (D3526)	0-6-0DM	1954
—	6	Peckett (2131)	0-4-0ST	1951
—	8	Barclay (885)	0-6-0ST	1900
—	322	Planet (3541)	4wDM	1952
Norma	3770	Hunslet (3770)	0-6-0ST	1952
—	—	Sentinel (9374)	4wVBT	1947
—	—	Hibberd (3057)	4wDM	1946
—	—	Planet	4wDM	1960

Stock

1 GWR auto-trailer
1 LMS brakevan
2 tank wagons
1 open wagon

Chasewater Light Railway SC

Founded in 1959 as the Railway Preservation Society (West Midlands District) the Group celebrated 30 years of existence in 1989. Activities now centre on 1-mile of track in Chasewater Pleasure Park
Location: Chasewater Pleasure Park, Brownhills, West Midlands
OS reference: SK 034070
Operating society/organisation: Chasewater Light Railway & Museum Co
Telephone: (0543) 452623
Car park: Pleasure park
Access by public transport: WMPTE 155 from Birmingham, 345 and 394 from Walsall. Midland Red and Chase Coaches 154 and 156 from Birmingham and Cannock
On site facilities: Refreshments (buffet coach), museum and souvenir shop. Extensive small relics collection. Lakeside walk
Catering facilities: Hot and cold buffet
Length of line: 1-mile, extension to 1¾ miles under construction
Public opening: Site open 11.00-dusk, Sundays only throughout the year. Trains 12.00-17.30 every 20min
Special events: 31 March, 1 April — Easter Holiday Steam; 21 April — Railwayana Fayre; 5/6 May — Mayday Kids Funday; 25/26 May — Spring Holiday; 16 June — Transport Scene Vintage Vehicle Rally; 1 July — Schools Day; 21 July — Best of British Motor Cycle Rally; 11 August — Military Day; 25/26 August — Bank Holiday Steam; 8 September — Chasewater Tank Engine Family; 13 October — Vintage & General Transport Rally; 3 November — Bonfire Steam; 15 December — Santa Specials; 29 December — Mince Pie Specials
Facilities for disabled: Access to station only, no steps, car park adjacent
Membership details: Brownhills West Station, Hednesford Road, Brownhills West, Walsall WS8 7LT

Industrial locomotives

Name	*No*	*Builder*	*Type*	*Built*
—	—	Peckett (917)	0-4-0ST	1902
Invicta	8	Barclay (2220)	0-4-0ST	1946
—	3	Barclay (1223)	0-4-0ST	1911
Lion	—	Peckett (1351)	0-4-0ST	1914
Alfred Paget	11	Neilson (2937)	0-4-0ST	1882
—	S100	H/Clarke (1822)	0-6-0T	1949
—	15	H/Clarke (431)	0-6-0ST	1895
Asbestos	4	H/Leslie (2780)	0-4-0ST	1909
—	21	Kent Cont (1612)	4wDM	1929
—	7	R/Hornsby (548641)	0-4-0DE	1961
—	—	Smiths	4wD crane	1951
—	59632	Sentinel (9632)	4wVBT	1957
—	1	M/Rail (1947)	4wPM	1919
—	61/50011	Fowler (4220015)	0-4-0DM	1962

Stock

Passenger services provided with ex-BR DMU trailer cars.
A selection of freight vehicles is also housed at the site, also several pre-Grouping coaches.

Chatterley Whitfield Mining Museum — M

Location: Tunstall, Stoke-on-Trent, Staffs ST6 8UN
Operating society/organisation: Chatterley Whitfield Mining Museum Trust
Telephone: (0782) 813337
Car parks: Main car park has capacity for up to 50 coaches; other overflow car parks are also available
Access by public transport: Nearest British Rail station, Longport. By bus, PMT Services operate regular (10min intervals Monday to Saturday) routes between Hanley
Refreshment facilities: Pit Canteen, 100-seat capacity, open 10.00 to 16.30, seven days a week
Other on-site facilities: Underground tours, British Coal's National Collection, film shows, restored winding engines, toilets, museum shop, education services, mining picture gallery, vintage car workshops
Length of line: ½-mile; currently non-passenger carrying; operated as a colliery railway siding
Public opening: Site open 10.00 to 17.00, seven days per week, 51 weeks per year. Last pit tour daily at 16.00
Period of public operation: Locomotive in steam on most opening days
Special events: Various throughout the year including annual Steam Carnival
Facilities for disabled: Disabled toilet, surface collection building; with prior booking driving for the disabled utilising retired pit pony can be arranged

Locomotives

Name	*No*	*Origin*	*Class*	*Type*	*Built*
—	2	NSR	L	0-6-2T	1923

Industrial locomotives

Name	*No*	*Builder*	*Type*	*Built*
—	6	Robert Heath	0-4-0ST	1886
The Welshman	—	M/Wardle (1207)	0-6-0ST	1890
Joseph	—	Hunslet (3163)	0-6-0ST	1944
—	9	YEC (2521)	0-6-0ST	1952
Hem Heath	3D	Bagnall (3119)	0-6-0DM	1956
—	—	NBL (27876)	0-4-0DM	1959
—	LO52	YEC (2745)	0-6-0DE	1960
—	13D	T/Hill (181V)	6wDH	1967

Stock

3 2ft 6in gauge underground manrider trains
2 2ft 4in gauge underground manrider trains
1 2ft gauge underground manrider train
1 Taylor & Hubbard steam crane
1 25-ton brake van
Various coal wagons

Cheddleton Railway Centre — SC

A Victorian country station set in the attractive Staffordshire moorlands, situated adjacent to the River Churnet. Pleasant walks can be taken alongside the Caldon Canal, and the Flint Mill Museum is only ⅔-mile away
Location: Cheddleton station, near Leek, Staffordshire
OS reference: SJ 983519
Operating society/organisation: North Staffordshire Railway Company (1978) Ltd
Telephone: Churnetside (0538) 360522, Sundays and Bank Holidays only (10.00-17.00). (0782) 503458 (evenings only)
Car park: Riverside car park and picnic area opposite the station
Access by public transport: BR Stoke-on-trent (10 miles). A regular bus service operated by Proctors, PMT and Stevensons from Hanley, Longton and Leek to Cheddleton village

Locomotives

Name	*No*	*Origin*	*Class*	*Type*	*Built*
—	4422	LMS	4F	0-6-0	1927
—	80136	BR	4MT	2-6-4T	1956
—	D3420	BR	08	0-6-0DE	1957

Industrial locomotives

Name	*No*	*Builder*	*Type*	*Built*
Josiah Wedgwood	52	Hunslet (3777)	0-6-0ST	1952

Locomotive notes: Both ex-Barry locomotives, Nos 4422 and 80136 have made big steps forward in their restoration in 1990. The 4F's boiler has been steam-tested and replaced on the frames. Steaming is expected during 1991. 80136's boiler overhaul commenced in 1990, and side tanks and bunker were replaced on the frames.

Owners

4422 (the 4422 Locomotive Fund)
D3420 (the NSR Diesel Group)

Restored to working order during 1990 former LMS Class 8P No 46203 *Princess Margaret Rose* is seen in action as it speeds through Draycott with its second run on 2 June 1990. Based at the Midland Railway Centre, and now privately-owned No 46203 was on display at Butlin's Pwllheli camp for many years.
A. R. Butcher

Another restoration to main line condition during 1990 was GWR 'King' class 4-6-0 No 6024 *King Edward*. It is seen here leaving Wood End tunnel with the 12.59 Tyseley-Stratford working on 15 April 1990. No 6024 was based at the Birmingham Railway Museum for a short period prior to transferring to the Didcot Railway Centre. *Phil Chilton*

On site facilities: The station contains a refreshment room, souvenir shop and small relics museum. On open days visitors are allowed to visit the yard, signalbox and new locomotive display hall. Modern toilets are also on site
Length of line: 400yd
Public opening: Easter-September inclusive — Sundays and bank holiday Mondays, 11.00-17.30 June, July and August. Sundays October-March 12.00-17.00
Special events: 9 June — Model Railway Day; 29 September — Classic Car Day; 15, 22 December — Santa & Steam. Other steam open days and steaming of 4422 — see press
Facilities for disabled: Access to most of the site is possible by wheelchair. Train rides by arrangement
Membership details: Richard Waterhouse c/o above address
Special notes: Coach parties catered for by prior arrangement. Guided tours available for parties on request. The Railway Centre is adjacent to a BR freight line which is now mothballed pending a decision on its future. The NSR Co would like to re-open the line to passenger trains as the 'Churnet Valley Railway'

Stock
2 ex-BR Mk 1 CK coaches
1 ex-BR Mk 1 TSO coach
2 ex-BR brake suburban coaches
1 ex-NSR coach body
1 ex-LMS 6-wheel full brake
1 ex-LMS goods brake van
2 ex-LMS box vans
2 ex-BR box vans
2 ex-LMS 5-plank wagons
1 ex-LMS hopper wagon
1 Esso tank wagon
1 7-ton diesel rail-mounted crane

Below:
Restoration of a mainline locomotive is an expensive project and often help is sought from firms and companies to defray the costs. In this case, MTM Speciality Chemicals of Cheddleton have paid for the hire of a 30-ton capacity road crane used to lift the overhauled boiler of LMS 4F No 4422 back into its frames. It is expected that this locomotive will steam for the first time since 1965 in 1991 at Cheddleton, North Staffordshire, 12 April 1990. *Richard Waterhouse*

Crewe Heritage Centre SC

Location: Crewe Heritage Centre, Vernon Way, Crewe
OS reference: SJ 709552
Operating society/organisation: Crewe Heritage Centre Trust/Railside Ltd
Telephone: (0270) 212130
Car park: Town Centre, Forge Street, Oak Street
Access by public transport: BR Crewe
Refreshment facilities: Cafe in APT
On-site facilities: Souvenir shop, collector's corner
Public opening: Limited opening during 1991 due to building work, please see press for details or telephone above number
Facilities for disabled: Toilets
Membership details: Heritage Centre Supporters Association, c/o above address
Notes: Steam locomotives passed for use over BR tracks are often stabled between duties on the North Wales Coast Line

Locomotives

Name	*No*	*Origin*	*Class*	*Type*	*Built*
—	D120	BR	45	1Co-Co1	1961
—	D1842	BR	47	Co-Co	1965
—	D5233	BR	25	Bo-Bo	1963
—	D7523	BR	25	Bo-Bo	1965

Industrial locomotives

Name	*No*	*Builder*	*Type*	*Built*
Jane Derbyshire	2	Barclay (1969)	0-4-0ST	1929
—	—	K/Stuart (4388)	0-4-0ST	1926
Robert	—	H/Clarke (1752)	0-6-0T	1943
Joseph	—	Hunslet (3163)	0-6-0T	1944

Rolling stock
APT vehicle Nos 48103, 48106, 48602, 48603, 48606, 49002, 49002 in use as cafe.
1 BR Mk 1 BSK
2 BR brake vans

Dean Forest Railway SC

Steam train rides previously operating alongside the Lydney to Parkend line have been extended southwards to Lydney Lakeside
Location: Norchard Steam Centre on the B4234, ¾-mile off A48 at Lydney, Glos
OS reference: SO 629044
Operating society/organisation: Dean Forest Railway Society in conjunction with owning company, Forest of Dean Railway Ltd
Telephone: Dean (0594) 843423
Car park: Adequate for cars and coaches — no charge
Access by public transport: None
On site facilities: A shop and refreshments are available at Norchard along with a museum, photographic display, riverside walk, forest trail and picnic area
Catering facilities: Hot and cold snacks, ploughman's lunches etc on steam days. Parties catered for by appointment
Length of line: 1¼ miles
Public opening: Daily for static display — shop, museum, refreshments open every Saturday and Sunday 11.00-17.00 and weekdays Easter to October. Steam days: Good Friday and all Bank Holiday Saturdays, Sundays and Mondays (Christmas/New Year excepted). All Sundays April-September. Wednesdays in June and July, daily throughout August
Special events: Thomas the Tank Engine's Friends 22/23 June, 21/22 September; Preservation Rally 20 October; Santa Specials 24 November, 1, 8, 15, 22, 29 December
Facilities for disabled: Access to museum, toilets and train rides
Membership details: Mr P. Brammell, 4 Pool Gardens, Higham, Gloucester

Locomotives

Name	*No*	*Origin*	*Class*	*Type*	*Built*
—	4121	GWR	5101	2-6-2T	1937
Pitchford Hall	4953	GWR	'Hall'	4-6-0	1928
—	5521	GWR	4575	2-6-2T	1928
—	5541	GWR	4575	2-6-2T	1928
—	9681	GWR	5700	0-6-0PT	1949
—	03.062	BR	03	0-6-0DM	1958
—	03.119	BR	03	0-6-0DM	1958
—	08.377	BR	08	0-6-0DE	1957

Industrial locomotives

Name	*No*	*Builder*	*Type*	*Built*
—	—	Barclay (2221)	0-4-0ST	1946
Jessie	—	Hunslet (1873)	0-6-0ST	1937
Uskmouth No 1	—	Peckett (2147)	0-4-0ST	1952
Wilbert	—	Hunslet (3806)	0-6-0ST	1953
Warrior	—	Hunslet (3823)	0-6-0ST	1954
—	—	Hunslet (2145)	0-4-0DM	1940
—	—	Fowler (4210101)	0-4-0DM	1955
—	—	Fowler (4210127)	0-4-0DM	1957
—	—	Hibberd (3947)	4wPM	1960
Cabot	39	R/Royce (10218)	0-6-0DH	1965
Lord Marshall of Goring	392	A/Barclay (392)	0-4-0DM	1954

Stock
8 ex-GWR coaches
16 ex-BR coaches
3 Wickham trolleys
1 steam crane Thos Smith (Rodley) TS 5027 (10 ton)
1 45 ton steam crane (Cowans Sheldon) ADRC 95222
50+ wagons

Below:
Pannier tank No 9681 trundles northwards through the Forest of Dean with its passenger train on 27 June 1990.
Hugh Ballantyne

Two of Glasgow Museum of Transport's locomotives are seen here in the new exhibition hall. They are GNSR 4-4-0 *Gordon Highlander* and HR 4-6-0 No 103. *G. J. C. Reid*

Foxfield Steam Railway

The railway, built in 1893 to connect a colliery to the national system, closed in 1965, has been re-opened.
Headquarters: Foxfield Steam Railway, Blythe Bridge, Stoke on Trent
Telephone: (0782) 396210 (weekends), 314532 (weekdays)
Main station: Blythe Bridge (Caverswall Road)
OS reference: SJ 957421
Car park: Blythe Bridge
Access by public transport: BR Blythe Bridge (400yd). PMT bus service to Blythe Bridge
Refreshment facilities: Blythe Bridge
Souvenir shop: Blythe Bridge
Passenger trains: Steam-hauled trains operate from Blythe Bridge (Caverswall Road) to Dilhorne Park and return
Length of line: 2¾ miles
Period of public operation: Steam trains operate Sundays and Bank Holidays only, April-September inclusive between Blythe Bridge and Dilhorne Park.
Special events: Santa Specials, weekends in December (advanced booking essential)
Facilities for disabled: Access to majority of facilities is on the level. For special requirements prior notice is desirable
Special notes: Wine and Dine Specials will operate on a number of Saturday evenings during the operating season, prior booking essential

Industrial locomotives

Name	*No*	*Builder*	*Type*	*Built*
Whiston	—	Hunslet (3694)	0-6-0ST	1950
Wimblebury	—	Hunslet (3839)	0-6-0ST	1956
Roker	—	RSH (7006)	0-4-0CT	1940
Cranford	—	Avonside (1919)	0-6-0ST	1924
Lewisham	—	Bagnall (2221)	0-6-0ST	1927
Hawarden	—	Bagnall (2623)	0-4-0ST	1940
Wolstanton No 3	—	Bagnall (3150)	0-6-0DM	1960
Bagnall	—	Bagnall (3207)	0-4-0DH	1961
Little Barford	—	Barclay (2069)	0-4-0ST	1939
—	—	E/Electric (788)	4wBE	1930
Spondon No 2	—	E/Electric (1130)	4wBE	1939
Henry Cort	—	Peckett (933)	0-4-0ST	1903
Ironbridge No 1	—	Peckett (1803)	0-4-0ST	1933
—	11	Peckett (2081)	0-4-0ST	1947
C.P.C.	—	Barclay (1964)	0-4-0ST	1929
Moss Bay	—	K/Stuart (4167)	0-4-0ST	1920
Rom River	—	K/Stuart (4421)	6wDM	1929
—	1	Barclay (1984)	0-4-0F	1930
Meaford No 4	—	Barclay (486)	0-6-0DH	1964
Helen	—	Simplex (2262)	4wDM	1924
Coronation	—	N/British (27097)	0-4-0DH	1953
—	—	R/Hornsby (395305)	0-4-0DM	1956
Gas-oil	—	R/Hornsby (408496)	0-4-0DM	1957
Hercules	—	Ruston (242915)	4wDM	1946
—	—	Thomas Hill (103C)	0-4-0DH	1957

Stock
5 coaches
4 scenery vans (some converted for other uses)
19 wagons
1 rail-mounted self-propelled diesel-electric crane

Ironbridge Gorge Museum

M

The railway items form only a small part of the displays on two of the museum's main sites, Blists Hill and Coalbrookdale. The Blists Hill site offers an opportunity to see a number of industrial and other activities being operated in meticulously reconstructed period buildings. A working foundry and ironworks are just two of the exciting exhibits. The Ironbridge Gorge was designated a World Heritage Site in 1987
Location: Ironbridge, Shropshire
OS reference: SJ 694033

Industrial locomotives

Name	*No*	*Builder*	*Type*	*Built*
—	—	Sentinel/Coalbrookdale (6185)	0-4-0VBT	1925
—	—	Sentinel/M/Wardle (6155)	0-4-0VBT	1925
—	5	Coalbrookdale	0-4-0ST	1865

All locomotives are at the Museum of Iron, Coalbrookdale

Note: A full-size working replica of Richard Trevithick's 1802 plateway locomotive, built by the Coalbrookdale Co, will be operating outside the Museum of Iron throughout the 1991 season. For further details please telephone the number above, or Telford (0952) 432751,

Operating society/organisation: Ironbridge Gorge Museum Trust, Ironbridge, Telford, Shropshire TF8 7AW
Telephone: Telford (0952) 433522
Car park: At the sites
Access by public transport: Various private bus companies, including Midland Red, Williamson's Shearings, Elcocks, Boultons.

On Sundays in the summer British Rail run a special Ironbridge Express from Birmingham via Wolverhampton to the Museum of Iron (for further details contact the Museum on the above telephone number).

A frequent 'Park & Ride' bus service operates on weekdays during school holidays, and on weekends from Easter to September. Vintage vehicles are used at weekends
Catering facilities: Licensed Victorian pub, sweet shop and tea rooms at the Blists Hill site, serving drinks and mainly cold snacks. Tea, coffee and light refreshments at the Museum of Iron and Rosehill House, Coalbrookdale
Public opening: Main sites including Museum of Iron and on Blists Hill, daily (except Christmas Eve and Christmas Day) 10.00-17.00, 10.00-18.00 during British Summer Time
Special notes: Tickets for all the sites or just for single sites available

Moseley Railway Museum SC

The collection concentrates on diesel and petrol power with some battery units. There are also hand, rope and chain operated vehicles on site. Emphasis is placed on operational locomotives and stock which, by prior appointment, can be observed and photographed performing the tasks for which they were originally designed. There are a number of unique running exhibits including the Kent petrol and Greenbat electric locomotives, and single cylinder Simplex and Ruston diesel units. Passenger train rides are always available on official open days
Location: Grounds of The Manor School, Northdowns Road, Cheadle, Cheshire SK8 5HA
Note: School will be renamed 'Margaret Danyers College' from 1 September 1991
OS reference: SJ 864871
Operating society: The Moseley Industrial Narrow Gauge Tramway Museum Society (MTM), address as above
Telephone: School hours 061-485 4372, any other time 061-485 4448, Colin Saxton
Car park: Extensive, free, adjacent to the museum buildings
Station: One boarding area adjacent to the museum. Other request stops
Length of line and gauge: Approximately ½-mile, 2ft gauge. Static lines in short lengths at other gauges
Access by public transport: Nearest BR station Cheadle Hulme (10min walk). Bus from Manchester 157, bus from Stockport S13, ask for Cheadle Adult Education Centre/Manor School
On site facilities: Free access and parking, a small selection of cold drinks and chocolate, access to all site
Period of public operation: Throughout the year. Second Sunday of every month plus any other time by appointment

Industrial locomotives

Name	*No*	*Builder*	*Type*	*Built*
—	—	G/Batley (2345)	4wBE	1951
—	—	G/Batley (2960)	4wBE	1959
—	—	G/Batley (420172)	4wBE	1969
—	—	H/Hunslet (6299)	4wDM	1964
—	—	Hunslet (4758)	4wDM	1954
—	—	Kent	4wDM	c1926
Ald Hague	—	Hiberd (3465)	4wDM	1954
—	—	Lister (8022)	4wDM	1936
—	—	Lister (38296)	4wDM	1952
—	—	Lister (52031)	4wDM	1960
—	—	M/Rail (4565)	4wDM	1928
—	—	M/Rail (7512)	4wDM	1938
—	—	M/Rail (7552)	4wDM	1948
—	—	M/Rail (8663)	4wDM	1941
—	—	M/Rail (8669)	4wDM	1941
—	—	M/Rail (8878)	4wDM	1944
The Lady D	—	M/Rail (8934)	4wDM	1944
Nick the Greek	—	M/Rail (8937)	4wDM	1944
—	—	M/Rail (9104)	4wDM	1942
—	—	M/Rail (11142)	4wDM	1960
Knothole Worker	—	M/Rail (22045)	4wDM	1959
Moseley	—	R/Hornsby (177639)	4wDM	1936
—	—	R/Hornsby (198278)	4wDM	1940
—	—	R/Hornsby (223667)	4wDM	1943
—	—	R/Hornsby (279647)	4wDM	1944
Neath Abbey	—	R/Hornsby (476106)	4wDM	1964
—	—	R/Hornsby	4wDM	—

Stock

Nearly 100 items of rolling stock, gauges from 10¼in to 2ft
3 Wickham Target Trolleys

Train services: Return journey (round trip) approximately 20min
Facilities for the disabled: The museum operates policy as Stockport Education Authority. Every effort is made to accommodate disabled persons who may find a visit easier during school hours when the main school premises are open
Special notes: Part of the site, including museum buildings and displays are now open
Membership details: Colin Saxton, Member of Staff in charge. School address or 60 Pingate Lane, Cheadle Hulme, Cheshire SK8 7LT or John Rowlands, Secretary/Treasurer School address

Severn Valley Railway

The railway hosts more main line engines than any other preserved line in the country, enjoying the backup of a large volunteer and professional workforce and extensive engineering workshops and equipment. Railway travel like it used to be
Headquarters: The Railway Station, Bewdley, Worcestershire
Telephone: Bewdley (0299) 403816; 24hr timetable — Bewdley (0299) 401001
Main stations: Bridgnorth, Bewdley, Kidderminster Town
Other public stations: Arley, Highley, Hampton Loade, Northwood Halt
OS reference: Bridgnorth SO 715926, Bewdley SO 793753
Car parks: At all stations
Access by public transport: Midland Red bus service X92 to Kidderminster and Bewdley and 890 to Bridgnorth. BR Sprinter service to Kidderminster (BR) with immediate connections to SVR station. Through tickets available from a wide range of BR stations
Refreshment facilities: Bridgnorth, Hampton Loade, Bewdley, Arley, Highley, Kidderminster Town and on most trains. Fully licensed bars at Bridgnorth and Kidderminster Town
Souvenir shops: Bridgnorth, Bewdley, Kidderminster Town and Model Railway at Bewdley
Depots: Bridgnorth (locomotives), Bewdley (stock)
Length of line: 16¼ miles
Passenger trains: Steam-hauled trains running frequently from Kidderminster Town to Bewdley and Bridgnorth. Diesel-hauled service on limited occasions
Period of public operation: Weekends March-October and Santa Steam specials in late November and December. Daily service mid-May to early October and all public Bank Holidays. Open for limited viewing at other times
Special events: Steam Enthusiasts' weekends in April and September; Diesel Gala weekends in May and October; Santa Specials end November and December; Mince Pie Specials in December/early January 1992
Facilities for disabled: Facilities available, special vehicle available to carry wheelchairs
Special notes: A number of special enthusiasts' weekends and special events are held when extra trains are operated. In addition supplementary

Locomotives

Name	*No*	*Origin*	*Class*	*Type*	*Built*
The Great Marquess	3442	LNER	K4	2-6-0	1938
Gordon	AD600	LMR	WD	2-10-0	1943
—	43106	LMS	4MT	2-6-0	1951
—	46443	LMS	2MT	2-6-0	1950
—	46521	LMS	2MT	2-6-0	1953
RAF Biggin Hill	45110	LMS	5MT	4-6-0	1935
Leander	5690	LMS	'Jubilee'	4-6-0	1936
Galatea	5699	LMS	'Jubilee'	4-6-0	1936
—	47383	LMS	3F	0-6-0T	1926
—	8233	LMS	8F	2-8-0	1940
—	2968	LMS	5P4F	2-6-0	1933
—	813	GWR	—	0-6-0ST	1901
—	2857	GWR	2800	2-8-0	1918
—	5164	GWR	5101	2-6-2T	1930
—	4150	GWR	5101	2-6-2T	1947
—	5764	GWR	5700	0-6-0PT	1929
—	7714	GWR	5700	0-6-0PT	1930
Raveningham Hall	6960	GWR	'Hall'	4-6-0	1944
—	4566	GWR	4500	2-6-2T	1924
Bradley Manor	7802	GWR	'Manor'	4-6-0	1939
*Erlestoke Manor**	7812	GWR	'Manor'	4-6-0	1939
Hinton Manor	7819	GWR	'Manor'	4-6-0	1939
Hagley Hall	4930	GWR	'Hall'	4-6-0	1929
—	1501	GWR	1500	0-6-0PT	1949
—	9303	GWR	4300	2-6-0	1932
Taw Valley	34027	SR	WC	4-6-2	1946
—	80079	BR	4MT	2-6-4T	1954
—	78019	BR	2MT	2-6-0	1954
—	75069	BR	4MT	4-6-0	1955
Western Ranger	D1013	BR	52	C-C	1962
Western Courier	D1062	BR	52	C-C	1963
—	D3022	BR	08	0-6-0DE	1952
—	D3586	BR	08	0-6-0DE	1953
—	12099	BR	11	0-6-0DE	1952
—	D7633	BR	25	Bo-Bo	1965

*Undergoing overhaul at Swindon Heritage Centre

Industrial locomotives

Name	*No*	*Builder*	*Type*	*Built*
—	4	Peckett (1738)	0-4-0ST	1928
Warwickshire	—	M/Wardle (2047)	0-6-0ST	1926
The Lady Armaghdale	—	Hunslet (686)	0-6-0T	1898
—	—	Ruston (319290)	0-4-0DM	1953
Alan	—	R/Hornsby (414304)	0-4-0DM	1957
William	—	R/Hornsby (408297)	0-4-0DM	1957

Stock
27 ex-GWR coaches
13 ex-LMS coaches
24 ex-BR Mk 1 coaches
8 ex-LNER coaches
Numerous examples of ex-GWR, LMS and other freight vehicles and two 30-ton steam cranes

Owners
3442 (family of the late Earl of Lindsay)
AD600 (The Royal Corps of Transport Museum Trustees)
43106 (the Ivatt 4 Fund)
46443 (SVR 46443 Fund)

trains with diesel haulage may be run on special occasions. 'Severn Valley Limited' Restaurant Car service operates on Sundays and dining car service on selected Saturday evenings. Advanced booking recommended
Membership details: Mrs Gwen Bailey c/o above address
Share details: Mr M. J. Draper, c/o above address
Feature article references: RW June 1989, p364 (Bewdley — pre-preservation); RW August 1990, p462 (No 8233)

Below:
Ivatt Class 2MT No 46443 heads a freight during an SVR gala event. This locomotive is expected to see service at Llangollen during the summer of 1991. *Martin W. Davies*

47383 (the Manchester Rail Travel Society)
8233 (the Stanier 8F Locomotive Society)
2968 (the Stanier Mogul Fund)
813 (the GWR 813 Fund)
2857 (the 2857 Fund)
5164 (The 51XX Fund)
4150 (the 4150 Locomotive Fund)
5764, 7714 (the Pannier Tank Fund)
4566 (the 4566 Fund)
7802, 7812 (the Erlestoke Manor Fund)
7819 (the Hinton Manor Fund)
1501 (the 15XX Trust)
9303 (the Great Western (SVR) Association)
80079 (the Passenger Tank Fund)
75069 (the 75069 Fund)
6960, 34027, 46521, 78019, D1013 (private)
D1062 (the Western Locomotive Association)
D3022 (the Class 08 Society)
D7633 (the SVR P/W Fund)

Telford Horsehay Steam Trust SC

Location: Horsehay, Telford, Shropshire
OS reference: SJ 675073
Operating society/organisation: Telford Horsehay Steam Trust, The Old Loco Shed, Horsehay, Telford, TF4 2LT
Public opening: Please contact for details

Locomotives

Name	*No*	*Origin*	*Class*	*Type*	*Built*
—	5619*	GWR	5600	0-6-2T	1925

*On loan to Swanage Railway until May 1991

Industrial locomotives

Name	*No*	*Builder*	*Type*	*Built*
Peter	—	Barclay (782)	0-6-0ST	1896
Tom	—	N/British (27414)	0-4-0DH	1954
Ironbridge No 3	—	Peckett (1990)	0-4-0ST	1940
—	—	Sentinel (9535)	4wVBT	1952
—	D2959	R/Hornsby (382824)	4wDM	1955

Stock
1 ex-BR Mk 1 coach
1 ex-GWR auto-trailer
1 ex-GWR Toad brake van
1 ex-GWR 3-ton hand crane
1 Wickham trolley
Various wagons

East Midlands

The Battlefield Steam Railway

A quiet country railway operated by members of the Shackerstone Railway Society Ltd, who are undertaking the extension of the line to Shenton and the battlefield of Bosworth
Headquarters: Market Bosworth Light Railway, Shackerstone station, near Market Bosworth, Leicestershire
Telephone: (0827) 715790 (weekends only); (0827) 880754 (weekdays and weekends only)
Main station: Shackerstone
Other public station: Market Bosworth
OS reference: SK 379066
Car park: Shackerstone
Access by public transport: Leicester City Bus, Lex Gibson Bros service 123 (tel: 24326). Leicester (Horsefair St) to Market Bosworth (square) weekdays and Sundays
Refreshment facilities: Shackerstone cafe on station. Licensed bar on train
Souvenir shop: Shackerstone
Museum: Shackerstone
Depot: Shackerstone
Length of line: 2¾ miles
Passenger trains: Shackerstone-Market Bosworth
Period of public operation: Passenger steam service will operate: Sundays and bank holiday Mondays 31 March-27 October. Station and museum only are open weekends throughout the year
Special events: 21 April — Friends of Thomas the Tank; 5/6 May — Teddy Bears' Picnic; 16 June — Fathers' Day; 7 July — Turtles Treat; 28 July — Car Boot; 11 August — Friends of Thomas the Tank; 8 September — Postman Pat Day; 22 September — Grandparents' Day; 2 November — Bonfire Special. Santa Specials 1, 7/8, 14/15, 21/22 December
Special notes: An extension of the line from Market Bosworth to Shenton and the battlefield of Bosworth is being undertaken and it is hoped to open in late 1991

Locomotives

Name	*No*	*Origin*	*Class*	*Type*	*Built*
—	D2245	BR	04	0-6-0DM	1956

Industrial locomotives

Name	*No*	*Builder*	*Type*	*Built*
Linda	—	Bagnall (2648)	0-4-0ST	1941
The King	—	Borrows (48)	0-4-0WT	1906
—	11	Hunslet (1493)	0-4-0ST	1925
Dunlop No 7	—	Peckett (2130)	0-4-0ST	1951
—	3	RSH (7537)	0-6-0T	1949
—	4	RSH (7684)	0-6-0T	1951
Lamport No 3	—	Bagnall (2670)	0-6-0ST	1942
Florence	2	Bagnall (3059)	0-6-0ST	1953
—	—*	R/Hornsby (235513)	4wDM	1945
—	—	R/Hornsby (263001)	4wDM	1949
The 1211 Squadron	—	R/Hornsby (347747)	0-6-0DM	1957
—	—	R/Hornsby (393304)	4wDM	1956
—	—	R/Hornsby (423657)	4wDM	1958
—	—	S/Crossley (7697)	0-6-0DM	1953

*Converted to steam power 1983

Stock
3 ex-BR Mk 1 coaches
2 ex-BR converted DMU coaches
3 parcels vans
2 rail mounted steam cranes
1 rail mounted diesel crane
Small number of wagons
ex-BR Griddle car
SR brake van

Great Central Railway

The original Great Central Railway's extension to London in 1899 was the last main line to be built in this country, most of which was closed in the 1960s. Steam-hauled services are now operating again through

Locomotives

Name	*No*	*Origin*	*Class*	*Type*	*Built*
Butler Henderson	506	GCR	'Director'	4-4-0	1919
—	68088	NER	Y7	0-4-0T	1923
—	5224	GWR	5205	2-8-0T	1924
Witherslack Hall	6990	GWR	'Hall'	4-6-0	1948

Above:
The Great Central Railway crosses Swithland reservoir; here ex-GWR 0-6-0PT No 7760, on loan from the Birmingham Railway Museum at Tyseley, is seen in action. *W. A. Sharman*

attractive rolling Leicestershire countryside crossing the picturesque Swithland reservoir. A project has now commenced to lay double track between Rothley and Loughborough. The collection of locomotives, includes the only surviving GCR express locomotive — *Butler Henderson*
Headquarters: Great Central Railway, (1976) PLC (Main Line Steam Trust Ltd), Loughborough Central Station, Great Central Road, Loughborough, Leicestershire LE11 1RW
Telephone: Loughborough (0509) 230726
Main station: Loughborough Central
Other public stations: Quorn & Woodhouse, Rothley
OS reference: SK 543194
Car park: Quorn, Rothley
Access by public transport: Loughborough BR station (¾-mile). Trent, South Notts, L'boro Coach & Bus Co and Midland Fox bus services to Loughborough bus station (¾-mile)
Refreshment facilities: Licensed buffet car and light refreshments on most trains. Sunday lunches on 13.00 train 'The Carillon'. Dinner train 'The Charnwood Forester' runs on certain Saturday nights, please contact railway for dates and reservations (advance booking only). Private charter trains available on request. Refreshment buffets at Loughborough and Leicester North
Souvenir shop: Loughborough
Museum: Loughborough
Depot: Loughborough
Length of line: 7½ miles (Leicester extension opened in November 1990)
Passenger trains: Loughborough-Leicester North
Period of public operation: Weekends throughout the year and Bank Holiday Mondays. Midweek services on certain Tuesdays,

Kolhapur	5593	LMS	'Jubilee'	4-6-0	1934
—	47406	LMS	3F	0-6-0T	1926
—	48305	LMS	8F	2-8-0	1943
—	1264	LNER	B1	4-6-0	1947
Mayflower	1306	LNER	B1	4-6-0	1948
—	69523	LNER	N2	0-6-2T	1921
Boscastle	34039	SR	WC	4-6-2	1946
Canadian Pacific	35005	SR	MN	4-6-2	1941
Brocklebank Line	35025	SR	MN	4-6-2	1948
—	92212	BR	9F	2-10-0	1959
—	D3101	BR	08	0-6-0DE	1955
—	D4067	BR	10	0-6-0DE	1961
Royal Highland Fusilier	55019	BR	55	Co-Co	1961

3-car DMU, Class 120 M59276, Class 127 M51616 and M51622

Wednesdays and Thursdays in June, July and August
Special events: Easter Fair; Thomas the Tank weekend, Bonfire Special and Santa trains planned
Facilities for disabled: Special carriage for wheelchair/disabled persons (advance notice required). Wheelchair access good at Quorn and Rothley, can be arranged at Loughborough with advance notification. Boarding ramps at all stations
Membership details: *Share enquiries:* Company Secretary, Great Central Railway (1976) PLC *Membership:* Membership Secretary, Main Line Steam Trust PLC. Both c/o above address
Feature article reference: RW August 1989, p493 (Locomotive 35025); RW June 1990, p349 (photo feature)

Industrial locomotives

Name	*No*	*Builder*	*Type*	*Built*
—	7597	RSH (7597)	0-6-0T	1949
—	68009	Hunslet (3825)	0-6-0ST	1953
Qwag	1	Ruston (371971)	4wDM	1954
Arthur Wright	D4279	Fowler (4210079)	0-4-0DE	1952
—	28	A/Barclay (400)	0-4-0DM	1956

Owners
506 (on loan from the National Railway Museum)
68088 (the Y7 Locomotive Society)
69523 (the Gresley Society)
6990 (the Witherslack Hall Locomotive Society)
61264 (the Thompson B1 Locomotive Society)
92212 (the 92212 Holdings Ltd)
7597 (Railway Vehicle Preservation Society)
55019 (the Deltic Preservation Society)
5593 (on loan from the Birmingham Railway Museum)

Leicestershire Museum M

Narrow gauge site railway (1ft 11½in gauge) formerly part of sewage pumping station that now forms museum site. Railway relaid in concrete by MSC scheme during early 1980s to original track layout. New track layout as an extension to original laid with 35lb rail on wooden sleepers. All the railway system is now run by volunteers. Original Simplex locomotive kept on site (non-operational at present). A Planet petrol locomotive and a Ruston diesel are used to demonstrate railway with typical tipper wagons and mine tubs. Line originally used for transferring solid material from screens to tip (about 100yd).
Location: Museum of Technology, Abbey Pumping Station, Corporation Road, off Abbey Lane, Leicester LE4 5PW
Operating group: Leicestershire County Council Museum Department and Museum Volunteer Group
Telephone: (0533) 661330
Car park: Free on site
Access by public transport: BR Leicester (London Road). Leicester City Bus route 29 or 29A direct from station (alight at Beaumont Leys Lane)
Length of line/gauge: About 200yd and 1ft 11½in gauge. No public riding on line
Operation: On Special Event Days: 1991 dates are Sundays 14.00-17.30. 14 April-15 September, 15 December, also Saturdays when volunteers are working on line. At other times museum is open 10.00 to 17.30 Monday to Saturday and 14.00 to 17.30 Sunday
On site facilities: Museum/shop/toilets/car park. Refreshments only on Special Event Days
Facilities for disabled: Access to museum lower floor and grounds. Steps to Engine House and refreshments on event days
Volunteer contact: Mr N. Pell, 211 Broadgate Lane, Anstey, Leics
Museum contact: Mr R. Bracegirdle, Industrial Museum, Snibston Mine, Ashby Road, Coalville, Leicestershire LE6 2LN. Telephone: 0530 510851
Other attractions: Museum holds various other railway items and small exhibition on Leicester & Swannington Railway. Some items only viewable by appointment or on Special Event Days

Industrial locomotive stock list
Narrow gauge:

Name	*No*	*Builder*	*Type*	*Built*
—	—	Motor Rail (52600)	4wPM	1931
Planet	—	(1776)	4wPM	1931
Ruston	—	(223700)	4wPM	1941

All at museum in Leicester

Standard gauge:

Name	*No*	*Builder*	*Type*	*Built*
Mars II	—	RSH (7493)	0-4-0ST	1948
—	2	Barclay (1815)	0-4-0F	1924
—	—	Brush (314)*	0-4-0ST	1906
Cadley Hill No 1	—	Hunslet (3851)	0-6-0ST	1962

*Originally Powiesland & Mason Railway No 6 taken over by GWR in 1924, No 921.
Standard gauge locomotives are at Snibston Mine and not available for viewing.

Stock
Several narrow gauge wagons

Midland Railway Centre

This centre is being established as not only an operating railway, but also as a 57-acre museum site and a 35-acre country park which will make the centre unique. A wide variety of exhibits are being collected

Location: Midland Railway Centre, Butterley station, near Ripley, Derbyshire

OS reference: SK 403520

Operating society/organisation: Midland Railway Trust Ltd

Telephone: Ripley (0773) 747674/570140

Fax: (0773) 570271

Car park: Butterley station on A61 1-mile north of Ripley

On site facilities: Newly opened museum, award winning country park, souvenir shop

Refreshment facilities: Restaurants at Butterley and Butterley Park, bar on most trains and extensive Wine & Dine service, the 'Midlander' (details from above address)

Length of line: 3½ miles

Public opening: Every Sunday and Bank Holiday Weekend throughout the year. Every Saturday April-October and December. Daily 29 March-7 April, 25 May-2 June, 19-27 October; Tuesdays, Wednesdays, Thursdays 2 July-8 September; Fridays 26 July-6 September

Journey time: Approximately 1hr

Special events: 1991 marks 21 years of the Midland Railway project, with events most weekends: Railway Art Show 31 March/1 April; Diesel & Steam, 3rd weeekend, April-October; Road Transport Display, 2nd weekend, April-October; Friends of Thomas the Tank 25 May-2 June, 19-27 October; Railway Industry Fair 6/7 July; Major Art Show 24-26 August; Santa Specials 30 November, 1, 7/8, 14/15, 21-24 December

Facilities for disabled: Toilets, special coach, access to shop and cafeteria, special weekend

Feature article reference: RW October 1990, p605 (photo feature)

Membership details: J. E. Hett, at above address

Locomotives

Name	*No*	*Origin*	*Class*	*Type*	*Built*
Princess Margaret Rose	46203	LMS	8P	4-6-2	1936
—	44027	LMS	4F	0-6-0	1924
—	44932*	LMS	5MT	4-6-0	1945
—	47564	LMS	3F	0-6-0T	1928
—	47327	LMS	3F	0-6-0T	1926
—	47357	LMS	3F	0-6-0T	1926
—	47445	LMS	3F	0-6-0T	1927
—	9395	LNWR	7F	0-8-0	1921
—	158A	MR	—	2-4-0	1866
—	53809	S&DJR	7F	2-8-0	1925
—	73129	BR	5MT	4-6-0	1956
—	80080	BR	4MT	2-6-4T	1954
—	80098	BR	4MT	2-6-4T	1955
—	92212	BR	9F	2-10-0	1959
—	92219	BR	9F	2-10-0	1959
—	D2138	BR	03	0-6-0DM	1960
—	12077	BR	11	0-6-0DE	1950
—	24061	BR	24	Bo-Bo	1960
—	D7671	BR	25	Bo-Bo	1967
—	40012	BR	40	1Co-Co1	1959
Great Gable	D4	BR	44	1Co-Co1	1959
—	45133	BR	45/1	1Co-Co1	1961
Tulyar	55015	BR	55	Co-Co	1961

*Shares time between here and Steamtown, Carnforth

Locomotive notes: In service 44932, 46203, 47357, 53809, 80080, 48151, 12077, D4, 45133, 55015 and D7671. Under restoration 44027, 47327, 80098, 40012, D2138. Awaiting repairs or stored 47445, 73129. Boiler and frames only 47564. Static display 158A, 9395.

Right:
BR Standard Class 9F No 92219 shortly after arrival at the Midland Railway Centre on 18 November 1990. It joins sister engine No 92214 which is undergoing restoration.
Robin Stewart-Smith

Industrial locomotives

Name	No	Builder	Type	Built
Gladys	—	Markham (109)	0-4-0ST	1894
Stanton	24	Barclay (1875)	0-4-0CT	1925
Victory	—	Peckett (1547)	0-4-0ST	1917
Lytham No 1	—	Peckett (2111)	0-4-0ST	1949
—	4	N/Wilson (454)	0-4-0ST	1894
—	—	Sentinel (9370)	4wVBT	1947
Andy	—	Fowler (16038)	0-4-0DM	1923
—	RS9	M/Rail (2024)	0-4-0DM	1921
—	RS12	M/Rail (460)	0-4-0DM	1912
—	—	M/Rail (1930)	0-4-0PM	1919
Boots	2	Barclay (2008)	0-4-0F	1935
—	D2959	R/Hornsby (384139)	0-4-0DE	1955
*Handyman**	—	H/Clarke (573)	0-4-0ST	1900
Albert Fields	—	H/Clarke (D1114)	0-6-0DM	1958
—†	—	Deutz (10249)	4wDM	1932
Campbell Brick Works†	—	M/Rail (60S364)	4wDM	1968
—†	—	M/Rail (5906)	4wDM	1932
—†	—	M/Rail (11246)	4wDM	1963
Hucknall Colliery‡	1	Ruston (480080)	4wDM	1963
Hucknall Colliery‡	3	Ruston	4wDM	1963
Oddson†	—	Marshall	4wVBT	1970
—†	—	Hunslet (7178)	4wDH	1971

*3ft gauge
†2ft gauge
‡2ft 6in gauge

Locomotive notes: In service Sentinel 9370, Hucknall 3, D2959, Deutz 10249, *Albert Fields*, *Oddson* M/Rails 5906/60S364. Under restoration *Andy* RS12. Awaiting repairs or stored *Stanton*, RS9, Ruston 480080, Hunslet 7178. Static display *Gladys*, 4, Boots No 2, *Handyman*, Hucknell 1.

Victory and M/Rail (1930) on loan to Derby Industrial Museum

Stock

Numerous carriages, wagons and cranes including MR Royal saloon, MR six-wheeled coach, BR horsebox, LMS travelling Post Office, MR brake van, L&YR saloon — all on display in new museum. LD&ECR six-wheel coach should be restored and on display for 1991

Owners

158A, 9395, 44027 (on loan from the National Railway Museum)
53809 (the 13809 Preservation Group)
47357, 47327, 47445, 47564, 73129 (Derby City Council)
80080, 80098 (the 80080 Holdings Ltd)
55015 (the Deltic Preservation Society)
D4 (the Peak Locomotive Preservation Co Ltd)
D7671 (Derby Industrial Museum)

National Tramway Museum

A journey back in time — an experience of living transport history with vintage horse-drawn, steam and electric trams running through a recreated townscape of authentic buildings, stone setts, iron railings and historic street furniture. The heart of the Museum is its collection of over 40 vintage trams and you can enjoy the thrill of travelling on the scenic mile-long track

Locomotives

Name	No	Builder	Type	Built
—	—	B/Peacock (2464)	0-4-0VB tram loco	1885
—	—	E/Electric (717)	4wE	1927
Rupert	—	R/Hornsby (223741)	4wDM	1944
GMJ	—	R/Hornsby (326058)	4wDM	1952
—	—	R/Hornsby (373363)	4wDM	1954

Also some 40 trams (including examples from Czechoslovakia, Portugal, USA and South Africa), about a third of which have been restored to working order

Location: Crich, near Matlock, Derbyshire DE4 5DP
OS reference: SK 345549
Operating society/organisation: Tramway Museum Society
Telephone: 0773 852565
Car park: Site, coach parking also available
Access by public transport: Nearest stations: Cromford (BR) or Alfreton & Mansfield Parkway (BR) then by bus; or Whatstandwell (BR) and steep uphill walk
On site facilities: Souvenir shop, bookshop and picnic areas. 1-mile electric tramway. Tramway period street, depots, displays, exhibitions and video theatre
Refreshment facilities: Hot and cold snacks and meals
Public opening: Saturdays, Sundays and bank holidays, 29 March to 3 November, plus Mondays to Thursdays April to end September. Also open Fridays, 31 May and 26 July through to end August. Additional openings: daily during autumn half-term week. Open: 10.00am to 5.30pm (6.30pm Saturdays/Sundays/Bank Holidays)
Facilities for disabled: Access available for most of site, but most trams excluded
Special Notes: Crich houses the largest collection of preserved trams in Europe and has a 1-mile working tramway on which restored electric trams are regularly operated. Special events are arranged throughout the season. Part of tram line occupies route of narrow gauge mineral railway built by George Stephenson
Membership details: From above address

Northampton Steam Railway

Headquarters: Pitsford & Brampton Station
Location: About 5 miles north of Northampton, Pitsford road off A50 or A508
Operating society/organisation: Northampton Steam Railway Preservation Society
Telephone: (0604) 22709
Car parks: On site
Access by public transport: None
On site facilities: Bar and restaurant open daily from 11.00 in the old station master's house, called the 'pines'. Sales stand
Length of line: ¾-mile at present
Public opening: Possible opening for mid-1991, construction work underway. Open most Sundays for viewing. Several open days planned. For details of opening times, events and journey times, please see railway press
Membership details: Mr F. Collins, Membership Secretary, 2 Watson Road, Long Buckby, Northants NN6 7PS

Locomotives

Name	*No*	*Origin*	*Class*	*Type*	*Built*
—	3862	GWR	2884	2-8-0	1942
Castel Dinas Bran	25035	BR	25	Bo-Bo	1963
—	27056	BR	27	Bo-Bo	1962
The Royal Artilleryman	45118	BR	45	1Co-Co1	1962

Industrial locomotives

Name	*No*	*Builder*	*Type*	*Built*
Colwyn	45	Kitson (5470)	0-6-0ST	1933
Yvonne	2945	Cockerill (2945)	0-4-0VBT	1920
—	2104	Peckett (2104)	0-4-0ST	1948
Bunty	—	Fowler (4210018)	0-4-0DM	1950
—	1	R/Hornsby (275886)	4wDM	1949
Sir Gyles Isham	764	R/Hornsby (319286)	0-4-0DM	1953

Stock
1 ex-SR double-decker motor coach
1 Pullman car
1 BR Mk 1 FK
1 BR Mk 1 TSO
1 BR Mk 1 SK
1 SR CCT van
2 BR Suburban coaches (1 CL, 1 BS)
1 BR Mk 2 SO
1 BR Mk 1 BSK
2 GWR coaches
1 GWR Siphon
2 LMS coaches
1 BR full brake
1 BR GUV
A number of various wagon types

Owners
3862 (the LNWR Preservation Society)
Colwyn (the *Colwyn* Preservation Society)

Northamptonshire Ironstone Railway Trust SC

Location: Hunsbury Hill Industrial Museum. Hunsbury Hill Country Park, Hunsbury Hill Road, Camp Hill, Northampton
OS reference: SP 735584
Operating organisation: Northamptonshire Ironstone Railway Trust Ltd, c/o Mr B. Denny, 5 Forest Road, Far Cotton, Northants NN4 9NZ

Industrial locomotives

Name	*No*	*Builder*	*Type*	*Built*
Vigilant	—	Hunslet (287)	0-4-0ST	1882
Brill	14	M/Wardle (1795)	0-4-0ST	1912
—	87	Peckett (2029)	0-6-0ST	1942*
Belvedere	—	Sentinel (9365)	0-4-0TG	1946
Musketeer	—	Sentinel (9369)	0-4-0TG	1946
Hylton	—	Planet (3967)	0-4-0DH	1961

Telephone: Northampton (0604) 767216
Access by public transport: Terminus of Northampton Transport bus routes, 24, 26 to Camp Hill from Northampton Greyfriars bus station. Operated by United Counties Bus Co on Sundays
On site facilities: Light refreshments, shop, toilets in car park. Brake van rides given on standard gauge railway. Children's playground and picnic areas within the park boundary
Length of line: 2¼ miles
Public opening: Easter Sunday to end of September and Bank Holidays
Times of opening: Museum: 11.00-17.00. Railway: First train 14.00, last train 17.10
Facilities for the disabled: Site relatively flat. Members willing to assist.

Spitfire 39	R/Hornsby (242868)	4wDM	1946
— 16	Hunslet (2087)	0-4-0DM	1940
— —	R/Hornsby (386875)	0-4-0DM	1955
— —	M/Rail (9711)	0-4-0DM	1946†
— —	M/Rail	0-4-0DM	1954†
— —	Lister (14006)	0-4-0PM	1950†

*Metre gauge
†2ft gauge

Special notes: Museum to the Ironstone Industry of Northamptonshire, the museum houses photographs, documents and other items connected with the Ironstone Industry of Northamptonshire. The railway has been relaid on part of the trackbed of the former 3ft 8½in gauge line of now defunct Hunsbury Ironstone Co
Membership details: Mrs J. Clayton, 9 High Street, Hallaton, Leicestershire

Peak Rail PLC SC

Locations: Buxton Midland Station Site, Buxton, Derbyshire; BR station, Matlock, Derbyshire; Darley Dale Station, Darley Dale, Derbyshire
OS reference: Buxton SK 296603, Matlock SK 060738
Operating society/organisation: Peak Rail PLC, Matlock Station, Matlock, Derbyshire
Car parks: On sites

Locomotives

Name	*No*	*Origin*	*Class*	*Type*	*Built*
—	48624	LMS	8F	2-8-0	1943
Penyghent	D8*	BR	44	1Co-Co1	1959
3rd Carabinier	D99*	BR	45	1Co-Co1	1961

Below:
This photograph shows the amount of work required to reopen a derelict railway. Vegetation has been cleared, trackbed graded and track laid, all to a standard to be approved by the Railway Inspectorate. *Tony Joyce*

Access by public transport: BR DMU service between Manchester Piccadilly and Buxton daily. Monday to Saturday between Derby and Matlock
Length of line: 2 miles between Matlock and Darley Dale (scheduled to open 1991)
On site facilities: Shop and buffet at Matlock
Public opening: Shop and buffet at Matlock open daily. Buxton site currently closed to public
Facilities for disabled: Advance notice not necessary but contact telephone is (0298) 79898
Special events: Santa Specials 1/2, 8/9, 15/16 December
Special notes: Peak Rail (Operations) Ltd, has outline permission to relay the 20-mile line from Buxton to Matlock from all relevant local authorities. At the time of writing track laying on the first 2 miles from Matlock to Darley Dale is almost complete and is scheduled to open in 1991

Sherwood Forester	D100*	BR	45	1Co-Co1	1961
—	D3429†	BR	08	0-6-0DE	1958
—	D5705*	BR	28	Co-Bo	1958
—	D7615†	BR	25	Bo-Bo	1963

*At Matlock
†At Darley Dale

Industrial locomotives

Name	*No*	*Builder*	*Type*	*Built*
Vulcan	—	V/Foundry (3272)	0-4-0ST	1918
Brookes No 1	—	Hunslet (2387)	0-6-0ST	1941
The Duke	2746	Bagnall (2746)	0-6-0ST	1944
*William**	—	Sentinel (9599)	0-4-0T	1956
—	E1	H/Clarke (D1199)	0-6-0DM	1960
—	1684	Hunslet (1684)	0-4-0T	1931
*Warrington**	150	RSH (7136)	0-6-0ST	1944
Meteor	7604	RSH (7604)	0-6-0T	1956
Moorbarrow	47	RSH (7849)	0-6-0T	1955
*Janine**	—	YEC	0-6-0DE	1963
—*	—	Simplex	0-4-0D	—
Robert	2068	Avonside (2068)	0-6-0ST	1933
*Cynthia**	—	R/Hornsby	0-4-0DM	—

*At Darley Dale

Owners
D8 (the North Notts Loco Group)
Vulcan (the Vulcan Loco Trust)

Rutland Railway Museum SC

This museum is dedicated to portraying the railway in industry, particularly iron ore mining, and has a wide range of industrial locomotives and rolling stock. Indeed, its collection of freight rolling stock is probably the most comprehensive in the country and regular demonstrations are a feature of the 'steam days'.
Location: Cottesmore Iron Ore Mines Siding, Ashwell Road, Cottesmore, near Oakham, Leicestershire — museum situated mid-way between villages of Cottesmore and Ashwell, approximately 4 miles north of Oakham (locally signposted)
OS reference: SK 886137
Operating society/organisation: Rutland Railway Museum, Cottesmore Iron Ore Mines Siding, Ashwell Road, Cottesmore, Nr Oakham, Leicestershire LE15 7BX
Telephone: Information, Stamford (0780) 63092/62384; Site, Oakham (0572) 813203
Car park: Free car park on site
Access by public transport: Nearest BR station, Oakham. Blands bus service, Leicester-Oakham-Cottesmore. Bartons buses, Nottingham-Melton Mowbray-Oakham-Ashwell, Corby/

Locomotives

Name	*No*	*Origin*	*Class*	*Type*	*Built*
NCB No 7	(D9518)	BR	14	0-6-0DH	1964
NCB No 3	(D9521)	BR	14	0-6-0DH	1964
BSC45	(D9520)	BR	14	0-6-0DH	1964
—	D9555	BR	14	0-6-0DH	1965

Industrial locomotives

Name	*No*	*Builder*	*Type*	*Built*
Albion	—	Barclay (776)	0-4-0ST	1896
Uppingham	—	Peckett (1257)	0-4-0ST	1912
Rhos	39	H/Clarke (1308)	0-6-0ST	1918
BSC No 2	—	Barclay (1931)	0-4-0ST	1927
Dora	—	Avonside (1973)	0-4-0ST	1927
Elizabeth	—	Peckett (1759)	0-4-0ST	1928
Singapore	Yard No 440	H/Leslie (3865)	0-4-0ST	1936
Swordfish	—	Barclay (2138)	0-6-0ST	1941
Drake	—	Barclay (2086)	0-4-0ST	1940
Sir Thomas Royden	—	Barclay (2088)	0-4-0ST	1940
Carlton No 3	—	Barclay (352)	0-4-0DM	1941
*Salmon**	8410/39	Barclay (2139)	0-6-0ST	1942
Coal Products No 6	—	Hunslet (2868) (Rebuilt Hunslet 3883 1963)	0-6-0ST	1943
—	8	Peckett (2110)	0-4-0ST	1950
—	3	N/British (27656)	0-4-0DH	1957
—	—	R/Hornsby (306092)	4wDM	1950
—	—	R/Hornsby (305302)	4wDM	1951
Phoenix	—	Hibberd (3887)	4wDM	1958
Stanton No 50	—	YEC (2670)	0-6-0DE	1958

Above:
***Dora* at the head of the two-brake van train at Cottesmore on 27 August 1990.**
Melvyn Hopwood

Peterborough-Oakham-Ashwell (services 117 and 125).
On site facilities: Free train rides, demonstration freight trains, refreshments, toilets, museum, shop, picnic sites, demonstration line with lineside walk and viewing areas, static displays of over 30 steam and diesel locomotives; over 70 wagons, vans and coaches (believed to be the largest collection of preserved freight stock in the UK)
Length of line: ¾-mile
Passenger trains: Regular shuttle service operates on open days using brake vans (approximately every 15min)
Public Opening: Open weekends or by arrangement, some weekdays (please telephone prior to visit). Regular steam days: 30*/31 March; 1 April; 5/6, 25*/26/27 May; 7 July; 4, 24-26 August; 29 September (*diesel service Saturday pm) Open 11.00-17.00
Special events: Children's Weekend 5/6 May; Teddy Bear Weekend 25*/26/27 May; Ironstone Day 7 July; Rutland Steam Gala 24/25/26 August (extra attractions)

Janus	No 28	YEC (2791)	0-6-0DE	1962
Colsterworth	1382	YEC (2872)	0-6-0DE	1962
—	65	Hunslet (3889)	0-6-0ST	1964
—	20-90-01	Barclay (499)	0-4-0DH	1965
—	20-90-02	R/Hornsby (504565)	0-4-0DH	1965
Betty	8411/04	R/Royce (10201)	0-4-0DH	1965
—	—	E/Electric (D1231)	0-6-0DH	1967
—	No 1	Hunslet (6688)	0-4-0DH	1968

*Currently on loan to British Steel Scunthorpe as 'Appleby Frodingham Steel Co' No 15

Locomotive notes: In service *Singapore*, *Dora*, *Salmon*, BSC45, NCB 3, D9555, *Colsterworth*, *Betty*, *Janus*, E/Electric D1231 and 1.

Stock
4 coaches
5 brake vans
12 covered goods vans
57 wagons (includes rakes of wagons as used in local ironstone and industrial railways)
2 rail cranes

Facilities for disabled: Site relatively flat. Members willing to assist.
Special notes: The museum houses an extensive collection of industrial locomotives and rolling stock typifying past activity in local ironstone quarries, nationwide mines and factories. A demonstration line approximately ¾-mile long has been relaid on the former MR Cottesmore mineral branch (originally built to tap local ironstone quarries), on which restored locomotives and stock are run. Among the latter is the only surviving Wisbech & Upwell Tramway coach and the last locomotive built for BR service at Swindon Works
Membership details: Membership Secretary, c/o above address

Eastern Counties

Bressingham Steam Museum SC

Five miles of various gauges of railway running through extensive gardens and a collection of well maintained and impressive main line locomotives. All the fun of the fair, with something for everyone including the mother who never wants to see another steam engine
Location: Two miles west of Diss on the A1066
OS reference: TM 080806
Operating society/organisation: Bressingham Steam Preservation Co Ltd, Bressingham Hall, Diss, Norfolk IP22 2AB
Telephone: Bressingham (037 988) 386 and 382
Car park: Steam Centre (free)
Access by public transport: Diss BR station (3 miles)
On site facilities: 9½/15/24in and standard gauge lines, totalling nearly 5 miles. Museum, steam roundabout, souvenir shop and restaurant, extensive gardens and plant centre
Public opening: Open daily from 28 April to 30 September. Special steam days every Sunday and Thursday, plus Wednesdays in July and August, and bank holiday Mondays, 10.00-17.30
Special events: Norfolk & Norwich Rover P4 Owners' Club Rally 26 May. 5th Annual Fire Rally 4 August
Facilities for disabled: Toilets, wheelchairs available. Able to take wheelchairs on Nursery Line Railway
Special notes: Reduced rates for coach parties. Prices on application
Feature article reference: RW June 1989, p338

Locomotives

Name	*No*	*Origin*	*Class*	*Type*	*Built*
Martello	662	LBSCR	A1X	0-6-0T	1875
Thundersley	80	LTSR	3P	4-4-2T	1909
Granville	102	LSWR	B4	0-4-0T	1893
—	490	GER	E4	2-4-0	1894
—	2500	LMS	4P	2-6-4T	1934
Royal Scot	6100	LMS	7P	4-6-0	1927
Duchess of Sutherland	6233	LMS	8P	4-6-2	1938
Oliver Cromwell	70013	BR	7MT	4-6-2	1951
Peer Gynt	5865	NSB	52	2-10-0	1944
Tom Paine	141R73	SNCF	141R	2-8-2	1948
King Haakon VII	377	NSB	21c	2-6-0	1919

Industrial locomotives

Name	*No*	*Builder*	*Type*	*Built*
Beckton	1	Neilson (4444)	0-4-0ST	1892
Beckton	25	Neilson (5087)	0-4-0ST	1896
William Francis	6841	B/Peacock (6841)	0-4-0+0-4-0T	1937
Millfield	—	RSH (7070)	0-4-0CT	1942
Bluebottle	—	Barclay (1472)	0-4-0F	1916

Narrow gauge locomotives

Gwynedd	—	Hunslet (316)	0-4-0ST	1883
George Sholto	—	Hunslet (994)	0-4-0ST	1909
Bronllwyd	—	H/Clarke (1643)	0-6-0WT	1930
Eigiau	—	O&K (5668)	0-4-0WT	1912
—	—	M/Rail (22120)	4wDM	1964
—	—	M/Rail (22253)	4wDM	1965

15in gauge locomotives

Rosenkavalier	—	Krupp (1662)	4-6-2	1937
Mannertreu	—	Krupp (1663)	4-6-2	1937
Flying Scotsman	4472	W. Stewart (4472)	4-6-2	1976

9½in gauge locomotives

Princess	—	Motor Gear & Engine Co	4-6-2	1947

Owners
80, 490, 2500, 70013 (on loan from the National Railway Museum)

Bure Valley Railway

Headquarters: Bure Valley Railway Ltd, Aylsham Station, Norwich Road, Aylsham, Norwich NR11 6BW
Telephone: (0263) 733858
Fax: (0263) 733814
Main public station: Aylsham (Norwich Road); Wroxham (Coltishall Road)
Other public stations: Coltishall, Brampton and Buxton
Car parks: 150 spaces at Aylsham, 60 spaces at Wroxham
OS reference: Aylsham — TG619540326456; Wroxham — TG630300318646
Access by public transport: BR Hoveton & Wroxham station (200yd)
Refreshment facilities: Aylsham and Wroxham
Journey time: Approximately 46min each way with 30min turnround
Length of line: 9 miles
Opening times: Daily Easter-end October; Weekends — November/December
Facilities for disabled: Toilets at Aylsham and Wroxham, no special rolling stock but each train can carry one wheelchair, advanced notice required
Special notes: Special parties and private charter trains by arrangement
Locomotives and carriages: On hire from RH&DR — *Black Prince*, *Winston Churchill* and *Samson*. Diesel hydraulic built by BVR. 20 fully enclosed coaches seating 20 people each

Colne Valley Railway

SC

An immaculate country station surrounded by green fields, the railway offers much to entertain as well as to educate. A well-deserved reputation for good food
Location: Castle Hedingham station, Yeldham Road, Castle Hedingham, Halstead, Essex CO9 3DZ
OS reference: TL 774362
Operating society/organisation: Colne Valley Railway Preservation Society
Telephone: Hedingham (0787) 61174
Car park: At the site (access from A604 road) between Castle Hedingham and Gt Yeldham
Access by public transport: Eastern National bus services No 88 Colchester-Halstead, 89 Halstead-Hedingham and Hedingham & District Nos 4 Braintree-Hedingham, 5 Sudbury-Hedingham. Nearest BR station — Braintree
On site facilities: Depot, museum, souvenir shop and restaurant, 4-acre riverside wooded picnic area, toilets, video carriage
Catering facilities: Restaurant carriage open daily 10.00-17.00 and 19.00-22.00 except 22 December-1 March and Mondays out of peak season. Pullman on train service on steam days for Sunday lunch, private hire and evening wine and dine also Sunday lunches and Saturday evening wine and dine Pullman service. *All must be prebooked.* Buffet carriage open most steam days also

Locomotives

Name	*No*	*Origin*	*Class*	*Type*	*Built*
—	D2041	BR	03	0-6-0DM	1959
—	03063	BR	03	0-6-0DM	1959
—	D2184	BR	03	0-6-0DM	1962
—	E79978	BR	—	AC Cars Railbus	1958

Industrial locomotives

Name	*No*	*Builder*	*Type*	*Built*
—	—	Barclay (349)	0-4-0DM	1941
Victory	—	Barclay (2199)	0-4-0ST	1945
—	190	Hunslet (3790)	0-6-0ST	1952
—	72	Vulcan (5309)	0-6-0ST	1945
Jupiter	—	RSH (7671)	0-6-0ST	1950
—	40	RSH (7765)	0-6-0T	1954
Barrington	—	Avonside (1875)	0-4-0ST	1921
—	1	H/Leslie (3715)	0-4-0ST	1928
—	YD43	R/Hornsby (221639)	0-4-0DM	1943
—	—	Hibbard (3147)	0-4-0DM	1947

Stock
11 ex-BR Mk 1 coaches
1 ex-Norwegian State Railway Class B 65 18803 (built 1926) Balcony Open Second (teak)
2 ex-Pullman cars, *Aquila* and *Hermione*
1 ex-BR Mk 1 (BG) exhibition carriage
1 ex-LNER BTO 43571 bar saloon
1 ex-LNER BG 41975
1 ex-GWR Engineers Observation Saloon
Steam crane DRG 80103
Sundry items of freight stock

Length of line: ¾-mile
Public opening: Daily for static displays except Christmas, New Year and January. Locomotives in steam most Sundays of the month between Easter and October inclusive, also every Bank Holiday and preceeding Sunday except Christmas and New Year. Schools' steam operation most weekdays in June and Santa Specials in December
Special events: Playgroup Specials, Rising Five Specials, Victorian Specials, Santa Specials, please telephone for details
Facilities for disabled: To look and ride only
Special notes: The railway has been completely rebuilt on part of the original Colne Valley line track-bed. The railway offers much of educational value specialising in school party visits by appointment at any time of the year. Catering approved by Egon Ronay 'Just-a-bite' guide, also new conference Pullman train now open for evening meals, Tuesdays-Fridays (static)
Membership details: P. Lemon, 78 Mountbatten Road, Braintree, Essex CM7 6TP

East Anglian Railway Museum SC

Location: Chappel & Wakes Colne station, near Colchester CO6 2DS
OS reference: TL 898289
Operating society/organisation: East Anglian Railway Museum, Chappel & Wakes Colne Station, Station Road, Wakes Colne, Essex CO6 2DS
Telephone: Colchester (0206) 242524
Car park: On site
Access by public transport: Chappel & Wakes Colne BR station. Also Eastern National/Hedingham Omnibus service No 88 Colchester-Halstead. Sundays Eastern National No 188 Colchester-Halstead
On site facilities: Light refreshments, bookshop, museum, souvenir shop, heritage centre and toilets
Public opening: Daily 09.00-17.30
Special events: Annual 'Camra' beer festival, early September. Santa Specials in December (please see press for details)
Special notes: Special steam days are held first Sunday of month March-October inclusive, plus Bank Holidays, Wednesdays in August. Three restored signalboxes, large goods shed and restoration shed. Victorian country junction station. Schools days and Santa steamings. Disabled visitors welcome — prior advice appreciated. Rail Riders sticker station.
Membership details: Membership Secretary, 24 Olivia Drive, Leigh-on-Sea, Essex SS9 3EG

Locomotives

Name	*No*	*Origin*	*Class*	*Type*	*Built*
—	69621	GER	N7	0-6-2T	1924
—	80151	BR	4MT	2-6-4T	1956
—	D2279	BR	04	0-6-0DM	1960

Industrial locomotives

Name	*No*	*Builder*	*Type*	*Built*
Jubilee	—	Bagnall (2542)	0-4-0ST	1936
—	1074	Barclay (1047)	0-4-0ST	1905
Belvoir	—	Barclay (2350)	0-6-0ST	1954
Gunby	68067	Hunslet (2413)	0-6-0ST	1941
—	2	M/Vick	0-4-0E	1912
—	1438	Peckett (1438)	0-4-0ST	1916
—	2039	Peckett (2039)	0-4-0ST	1943
Penn Green	54	RSH (7031)	0-6-0ST	1941
—	AMW144	Barclay (333)	0-4-0DM	1938
—	23	Fowler (4220039)	0-4-0DH	1965
—	2029	Simplex (2029)	0-4-0PM	1920

Locomotive notes: 69621 may operate on Swanage Railway, summer 1991.

Stock
3 ex-BR Mk 1 coaches
1 ex-BR sleeping coach
1 ex-BR Mk 1 full brake
1 ex-LNER Buffet car
1 LNER pigeon van
1 ex-LNER coach
1 GER 6-wheel full brake
1 GER 6-wheel family saloon
1 GER 4-wheel coach
1 ex-GER bogie coach
2 MSL 6-wheel coaches
1 SR PMV
1 ex-BR CCT
1 ex-BR 13-ton open wagon
1 Lomac wagon
1 ex-LMS 12-ton open wagon
2 Wickham Trolleys
1 GWR Toad Brakevan
1 ex-BR brake van

East Anglia Transport Museum

SC

The East Suffolk Light Railway is the title given to the 2ft gauge railway, which winds its way some 300yd or so along the northern perimeter of the museum site, between the stations of Chapel Road and Woodside. The railway commenced operation in 1973 and aims to re-create a typical passenger-carrying light railway of years gone by. Many aspects of railway interest can be found along its length. The track came from Leziate sand-quarry and Canvey island, as well as from the Southwold Railway. There is also a signalbox from the Lowestoft-Great Yarmouth line, and signals from various local locations; all of which help to set the overall scene

Location: Carlton Colville, three miles south-west of Lowestoft in Suffolk
OS reference: TM 505903
Operating society/organisation: East Anglia Transport Museum Society Ltd, Chapel Road, Carlton Colville, Lowestoft, Suffolk NR33 8BL
Telephone: (0502) 518459
Car park: On site

Industrial locomotives

Name	*No*	*Builder*	*Type*	*Built*
—	1	M/Rail (5902)	4wDM	1932
—	2	M/Rail (5912)	4wDM	1934
—	4	R/Hornsby (177604)	4wDM	1936

Locomotive notes: During the 1991 season, M/Rail (5912) is likely to be the sole source of motive-power. The other two locomotives are under course of restoration, as finance and time allows. The return journey takes 10 to 15min.

Stock
Locally designed and built covered coach
Small selection of wagons
Van body ex-Southwold Railway

Access by public transport: Services 111, 119, 631, 671 from Lowestoft. BR Oulton Broad South (1 mile)
On site facilities: Refreshments, souvenir and bookshop, toilets, working transport museum, including trams, narrow-gauge railway, trolleybuses, steam-rollers and other commercial and public transport vehicles
Period of public opening: Easter Sunday and Monday, and Sundays from beginning of May to end of September, also other Bank Holidays in this period; open from 11.00. Saturdays from the beginning of June to the end of September, and every weekday during August; open from 14.00
Special notes: Limited facilities for the disabled
Membership details: From the above address

Leighton Buzzard Railway

In places, this line bears the air of a Continental roadside tramway, running as it does behind the backs of houses before it passes into the open countryside; there is no denying the quaintness of its engines or the friendliness of the little trains and their staff

Headquarters: Leighton Buzzard Railway, Page's Park station, Billington Road, Leighton Buzzard LU7 8TN
OS reference: Page's Park SP 928242
Telephone: (0525) 373888, 24hr answerphone with service and event details
Main station: Pages Park (A4146, Leighton Buzzard)
Other public stations: Halts at Vandyke Road and Stonehenge Works
Car park: Pages Park
Access by public transport: Leighton Buzzard BR station then by bus to town centre
Refreshment facilities: Pages Park
Souvenir shop: Pages Park

Locomotives

Name	*No*	*Builder*	*Type*	*Built*
Chaloner	1	de Winton	0-4-0VBT	1877
Pixie	2	K/Stuart (4260)	0-4-0ST	1922
Rishra	3	Baguley (2007)	0-4-0T	1921
Doll	4	Barclay (1641)	0-6-0T	1919
Elf	5	O&K (12740)	0-6-0WT	1936
Caravan	6	Simplex (7129)	4wDM	1938
Falcon	7	O&K (8986)	4wDM	1939
Gollum	8	Ruston (217999)	4wDM	1942
Madge	9	O&K (7600)	4wDM	1935
Haydn Taylor	10	Simplex (7956)	4wDM	1945
P. C. Allen	11	O&K (5834)	0-4-0WT	1912
Carbon	12	Simplex (6012)	4wPM	1930
Arkle	13	M/Rail (7108)	4wDM	1937
—	14	Hunslet (3646)	4wDM	1946
Tom Bombadil	15	Hibberd (2514)	4wDM	1941
Thorin Oakenshield	16	Lister (11221)	4wDM	1939
Damredub	17	Simplex (7036)	4wDM	1936
Fëanor	18	M/Rail (11003)	4wDM	1956
—	19	M/Rail (11298)	4wDM	1965
—	20	M/Rail (60s317)	4wDM	1966
Festoon	21	Simplex (4570)	4wPM	1929
Fingolfin	22	under construction	—	—
—	23	Ruston (164346)	4wDM	1932
Ad-a-Cab	25	Simplex (7214)	4wDM	1938

Depots: Pages Park and Stonehenge Works
Length of line: 3 miles, 2ft gauge
Journey time: Single 25min, return 60min
Passenger trains: Pages Park-Stonehenge Works
Period of public operation: Sundays 24 March-6 October; Good Friday; Easter Saturday; Bank Holiday Mondays 1 April, 6/27 May, 26 August; Wednesdays 31 July-28 August; Thursdays 1-22 August
Special events: Summer Steam Gala 23 June; September Steam Up 7/8 September
Facilities for disabled: Access to shop, buffet, platform. Toilet facilities are specially available June-September. Wheelchairs can be conveyed on trains. Those persons who can be transferred from chair to train are conveyed by first available train
Membership details: Membership secretary c/o above address

Yimkin	26	Ruston (203026)	4wDM	1942
Poppy	27	Ruston (408430)	4wDM	1957
RAF Stanbridge	28	Ruston (200516)	4wDM	1949
Creepy	29	Hunslet (6008)	4wDM	1963
—	30	M/Rail (8695)	4wDM	1941
—	31	Lister (4228)	4wDM	1931
—	32	Ruston (172892)	4wDM	1934
—	33	Hibberd (3582)	4wDM	1954
—	34	Hunslet (2536)	4wDM	1941
—	35	Hunslet (6619)	0-4-0DM	1966
T. W. Lewis	39	Ruston (375316)	4wDM	1954
—	43	Simplex (10409)	4wDM	1954
Kestrel	44	Simplex (7933)	4wDM	1941

Dismantled for spares: M/Rail 4805, 5613, 5612, 5603; Ruston 218016
Converted to brake vans: M/Rail 5608, 5875

Locomotive notes: 18 of these locomotives operated during 1990 and steam locomotives 1, 3, 4 and 5 are potentially operational during 1991, though possibly only during Gala Days.

Stock
8 coaches and a number of miscellaneous vehicles

Right:
Part of the line-up of motive power at Page's Park station on the occasion of the September 1990 gala weekend. *Rishra*, *Peter Pan* (visitor), *Pixie*, *Chaloner* and *Doll* complete the line-up. *Robin Stewart-Smith*

Nene Valley Railway

This unique railway's collection includes locomotives and coaches from 10 countries and two continents. It is a regular location for TV and film makers — from films like *Octopussy* with Roger Moore as 007 to TV — *Hanney, Christobel* — to commercials for cars, beer and soft drinks. The railway and the pleasant Cambridgeshire countryside have doubled for as diverse locations as Russia and Spain. *Thomas* also makes regular appearances to delight the children
Headquarters: Nene Valley Railway, Wansford station, Stibbington, Peterborough, Cambs PE8 6LR

Locomotives

Name	*No*	*Origin*	*Class*	*Type*	*Built*
—	80.014	DB	80	0-6-0T	1927
—	996	DB	E	4-6-2	1950
—	656	DSB	F	0-6-0T	1949
Danish Seaways	740	DSB	S	2-6-4T	1928
—	101	SJ	B	4-6-0	1944
—	1178	SJ	S	2-6-2T	1914
—	3.628	Nord	3500	4-6-0	1911
92 Squadron	34081	SR	BB	4-6-2	1948
—	5231	BR	5MT	4-6-0	1936
City of Peterborough	73050	BR	5MT	4-6-0	1954
—	D9516	BR	14	0-6-0DM	1964
—	D9523	BR	14	0-6-0DM	1964
—	14029 (D9529)	BR	14	0-6-0DM	1964
Atlantic Conveyor	D306	BR	40	1Co-Co1	1960
—	64.305-6	DB	64	2-6-2T	1936

Telephone: Stamford (0780) 782854; talking timetable (0780) 782921
Main station: Wansford
Other public stations: Orton Mere, Ferry Meadows, Peterborough NVR ½-mile from city centre
OS reference: TL 903979
Car park: Wansford, Orton Mere, Ferry Meadows, Peterborough NVR
Access by public transport: Buses from Peterborough to Orton Mere and Ferry Meadows
Refreshment facilities: Wansford, (restaurant bar coach on platform 4). Orton Mere. Light snacks available. Bar coach on most trains
Souvenir shops: Wansford, Orton Mere, Ferry Meadows
Museum: Wansford
Depot: Wansford
Length of line: 7½ miles
Passenger trains: Yarwell Mill-Wansford-Orton Mere-Peterborough NVR
Period of public operation: Weekends April-end-October; Sundays March and November, bank holidays; Mid-week July and August. School specials 12-14 and 19-21 June and Bank Holidays, mid-week June, July and August
Special events: Thomas the Tank Weekends May, August and end October; Santa Specials in December (phone for details)
Facilities for disabled: Ramp access to all stations and shops. Toilets adjacent to Wansford station souvenir shop. Disabled persons and helpers are eligible for concessionary fares. Passengers can be assisted on and off trains

Industrial locomotives

Name	*No*	*Builder*	*Type*	*Built*
—	804	Alco (77778)	Bo-Bo	1949
—	—	Avonside (1945)	0-6-0ST	1926
—	90432	Barclay (2248)	0-4-0ST	1948
Toby	—	Cockerill (1626)	0-4-0VBT	1890
Muriel	—	E/Electric (1123)	0-4-0DH	1966
Ryan	—	H/Clarke (1539)	0-6-0ST	1924
Thomas	—	H/Clarke (1800)	0-6-0T	1947
Jacks Green	—	Hunslet (1953)	0-6-0ST	1939
—	68081	Hunslet (2855)	0-6-0ST	1943
Horsa	—	RSH (8368)	0-4-0DE	1961
—	—	R/Hornsby (294268)	4wDM	1951
—	—	R-Hornsby (321734)	4wDM	1949
Doncaster	—	YEC (2654)	0-4-0DE	1957

Stock
13 BR Mk 1 coaches
Wagons Lits sleeping car, Italian-built
Wagons Lits sleeping car, Belgian-built
6 coaches from Denmark
1 coach from France
1 coach from Norway
Leyland Vehicles Experimental coach
3 steam rail cranes
SR Travelling Post Office
Various items of freight stock

Owners
34081 (the Battle of Britain Locomotive Preservation Society)
73050 (Peterborough City Council)
3.628 (Science Museum)
996 and 804 (Museum of World Railways)

Membership details: Bill Foreman and Wendy Reed, 87 Flamborough Close, Woodstow, Peterborough PE2 9LP

North Norfolk Railway

Part of the former Midland & Great Northern Joint Railway, other elements of the LNER have crept in in the guise of the Great Eastern 'J15', the 'B12' now being restored and a variety of Gresley coaches. The newly opened extension to Holt has opened up extensive views of the Norfolk coast and the steep gradients belie the county's reputation for flatness
Headquarters: North Norfolk Railway, Sheringham station, Sheringham, Norfolk, NR26 8RA
Telephone: Sheringham (0263) 822045. Talking timetable: (0263) 825449
Main station: Sheringham
Other public stations: Weybourne, Kelling Heath Park, Holt

Locomotives

Name	*No*	*Origin*	*Class*	*Type*	*Built*
—	564	GER	J15	0-6-0	1912
—	61572	LNER	B12	4-6-0	1928
—	12131	BR	11	0-6-0DE	1952
—	27066	BR	27	Bo-Bo	1962

Industrial locomotives

Name	*No*	*Builder*	*Type*	*Built*
Pony	—	H/Leslie (2918)	0-4-0ST	1912
Ring Haw	—	Hunslet (1982)	0-6-0ST	1940
—	3809	Hunslet (3809)	0-6-0ST	1954
Harlaxton	—	Barclay (2107)	0-6-0T	1941
Birchenwood	4	Bagnall (2680)	0-6-0ST	1944
John D Hammer	5	Peckett (1970)	0-6-0ST	1939
—	12	RSH (7845)	0-6-0T	1955
—	10	E/Electric (C8431)	0-4-0DH	1963
—	4	Fowler (4210080)	4wDM	1953

OS reference: Sheringham TG 156430, Weybourne TG 118419
Car parks: Sheringham (old goods yard), Weybourne, Holt
Access by public transport: Sheringham BR station (200 yards)
Refreshment facilities: Sheringham, Weybourne
Souvenir shops: Sheringham, Weybourne
Museum: Sheringham
Depot: Weybourne
Length of line: 5¼ miles
Passenger trains: Steeply graded, Sheringham-Weybourne-Holt
Period of public operation: Please contact for details
Special events: Santa Specials in December
Facilities for disabled: Specially adapted Pullman Car available, advanced booking essential
Feature article reference: RW July 1989, p433 (last days of the M&GNJR)

Wissington	—	H/Clarke (1700)	0-6-0ST	1938
—	—	Bagnall (2370)	0-6-0F	1929
Edmundsons	—	Barclay (2168)	0-4-0ST	1940

Stock
2 Wagon und Maschinenbau ex-BR diesel railbuses
10 ex-BR coaches
2 ex-SR 'Brighton Belle' Pullman cars
7 ex-LNER coaches
Small number of wagons

Owners
564 and 61572 (the Midland & Great Northern Railway Society)

Below:
Sheringham station on the North Norfolk Railway on 28 May 1990 as *Ring Haw* prepares to leave with the 15.30 service to Holt.
John C. Baker

Wells & Walsingham Light Railway

One man's railway, the life and love of retired naval commander, Roy Francis, this delightful line which is totally uncommercialised runs along the old Wells branch to Walsingham where the old station has been transformed into a Russian Orthodox Church by the addition of an onion-shaped dome to its roof. A must if you find yourself nearby

Location: On A149, Stiffkey Road, Wells next the Sea, Norfolk
Operating organisation: Wells & Walsingham Light Railway
Car park: Yes
Access by public transport: Eastern Counties buses
On site facilities: Souvenir and tea shop
Length of line: 4 miles, 10¼in gauge

Public opening: Daily Easter to the end September
Special notes: Journey may be commenced at either end, believed to be world's longest 10¼in gauge line. Built on the old Wells & Fakenham Railway trackbed. Old Swainsthorpe signalbox on site at Wells. Motive power is provided by a Garratt locomotive

North West

Blackpool & Fleetwood Tramway

Operating organisation: Blackpool Transport Services Ltd, Rigby Road, Blackpool, Lancashire FY1 5DD
Telephone: (0253) 23931
Length of line: 11½ miles, standard gauge
Period of public operation: Daily throughout the year
Number of trams: 86 double- and single-deck trams
The Blackpool & Fleetwood Tramway is the sole surviving street tramway system in the United Kingdom and attracts visitors from all over the country. During the autumn the streets are illuminated and several specially decorated trams are used

East Lancashire Railway

A very popular yet newly opened railway by the East Lancs Railway Society in close co-operation with local authorities, the line won the 1987 ARPS award. Visit the line to find out the cause of the line's popularity and success
Location: Bolton Street Station, Bury, Lancashire BL9 0EY
OS reference: SD 803109
Operating society/organisation: East Lancashire Railway Preservation Society
Telephone: 061 764 7790
Access by public transport: BR to Bury Interchange (not Sundays). GM Buses, Rossendale Transport and Ribble Motor Services to Bury
On site facilities: Refreshments normally available when trains are running. Buffet car service on most trains. Souvenir shop, transport museum
Length of line: Approximately 4 miles

Right:
An example of the earliest BR-designed station to be preserved is Bury Bolton Street station. No 76079 prepares to leave for Ramsbottom with the 11.00 service on 6 May 1990.
Brian Dobbs

Locomotives

Name	*No*	*Origin*	*Class*	*Type*	*Built*
—	7229	GWR	7200	2-8-2T	1935
—	45337	LMS	5MT	4-6-0	1937
—	46428	LMS	2MT	2-6-0	1948
—	73156	BR	5MT	4-6-0	1956
—	76079	BR	4MT	2-6-0	1957
—	80097	BR	4MT	2-6-4T	1954
Morning Star	92207	BR	9F	2-10-0	1959
—	D335	BR	40	1Co-Co1	1961

Public opening: Steam and diesel-hauled services operate on Saturdays, Sundays and Bank Holidays throughout the year. Santa Specials (advanced booking only) in December
Special events: 26 August — Teddy Bear's Picnic; Santa Specials — weekends in December; Diesel Enthusiasts' Weekend and Friends of Thomas the Tank Engine Days — please apply for details; Irwell Valley Diner, Wine & Dine Trains (advance booking only — please apply for details)
Special Notes: The Society re-opened a section of the former Bury-Rawtenstall line as far as Ramsbottom, in July 1987 and carried over 26,000 passengers in its first three months of operation. The remaining 4½ miles is due to re-open in 1991
Feature article reference: RW May 1990, p269

—	40145	BR	40	1Co-Co1	1961
Onslaught	D832	BR	42	B-B	1961
Western Prince	D1041	BR	52	C-C	1962
—	D2587	BR	05	0-6-0DM	1959
—	D2767	BR	—	0-4-0DH	1960
—	D2774	BR	—	0-4-0DH	1960
—	D2956	BR	01	0-4-0DM	1956
—	D5054	BR	24	Bo-Bo	1960
—	D7076	BR	35	B-B	1963
—	25901	BR	25	Bo-Bo	1966
—	D7659	BR	25	Bo-Bo	1966
—	D9531	BR	14	0-6-0DH	1965

Industrial locomotives

Name	*No*	*Builder*	*Type*	*Built*
Gothenburg	32	H/Clarke (680)	0-6-0T	1903
Phoenix	70	H/Clarke (1464)	0-6-0T	1921
Sir Robert Peel	8	Hunslet (3789)	0-6-0ST	1953
Shropshire	193	Hunslet (3793)	0-6-0ST	1953
—	1	Barclay (1927)	0-4-0ST	1927
MEA No 1	1	RSH (7683)	0-6-0T	1951
—	DH16	Sentinel (10175)	4wDH	1964
Mr Mercury	1	Hibberd (3438)	4wDM	1950
Winfield	—	M/Rail (9009)	4wDM	1948

Stock
22 BR Mk 1 coaches
Battery-electric multiple-unit Nos SC79998/9
1 L&YR coach
2 Bogie Guards coaches
3 Newspaper vans
Cravens 50-ton steam crane RS1013/50 (1930), NER 5-ton hand crane DB915390 (1880) and Smiths 5-ton diesel crane (1939) plus a small number of other goods vehicles

Owners
40145 (the Class 40 Preservation Society)
Battery-electric MU (on loan from the West Yorkshire Transport Museum)

Museum of Science & Industry M

Location: Liverpool Road Station, Castlefield, Manchester (off Deansgate near Granada)
OS reference: SJ 831987
Operating society/organisation: Museum of Science & Industry, Liverpool Road Station, Liverpool Road, Castlefield, Manchester M3 4JP
Telephone: 061 832 2244
Car parks: On site, plus parking in the area (museum car park 50p)
Access by public transport: Manchester Victoria, Piccadilly and Oxford Road BR stations. GM bus 33
On site facilities: Oldest passenger railway station, weekend train rides, former railway warehouses containing exhibitions about science, industry,

Locomotives

Name	*No*	*Origin*	*Class*	*Type*	*Built*
Magpie	44806	LMS	5MT	4-6-0	1944
Pender	3*	IoMR	—	2-4-0T	1873
Novelty	Powered replica of 1929 locomotive using some original parts				
—	3157†	PR	—	4-4-0	1911
—	2352‡	SAR	GL	4-8-2+2-8-4	1929
Ariadne	1505 (27001)	BR	EM2 (77)	Co-Co	1954
Hector	26048	BR	EM1 (76)	Bo-Bo cab only	—

Industrial locomotives

Name	*No*	*Builder*	*Type*	*Built*
Agecroft No 3	—	RSH (7681)	0-4-0ST	1951
—	—	Barclay (1964)	0-4-0ST	1929
—	—	E/Electric (1378)	4wBE	1944

aviation, space, water supply and sewage disposal and electricity. New exhibitions include Xperiment!, the hands-on science centre and the 'out of this world' space gallery. Also a licensed buffet and shop
Public opening: Daily except 23, 24 and 25 December, including Saturdays and Sundays, 10.00-17.00. Entrance in Lower Byrom Street. Admission charged
Special events: Castlefield Carnival held during the August Bank Holiday weekend
Special notes: Some wheelchair access, toilets for the disabled, lecture and conference facilities

—	—	H/Leslie (3682)	Bo-Bo	1927
—	—	Fowler	0-6-0	—

*Ex-Isle of Man Railways, 3ft gauge, sectioned (B/Peacock 1255)
†Ex-Pakistan Railways, 5ft 6in gauge (V/Foundry 3064)
‡Ex-South African Railways, 3ft 6in gauge (B/Peacock 6693)

Rolling stock
BR Restaurant Car (RU(A), BRCW Co Ltd, 1959
31217 BR passenger bogie brake van, LMS (Wolverton), 1944
c1908 L&YR 3rd class carriage re-built 1923 as Medical Examination Car, LMS No 10825
B955209 20-ton goods brake van, BR (Ashford), 1962
B782903 4-wheeled covered goods van, BR (Wolverton), 1961
B783709 4-wheeled covered goods van, BR (Wolverton), 1962
3-plank loose coupled goods wagon, MSLR (Dukinfield)?, c1890

Owners
Novelty on loan from the Science Museum, London
RU (A) Birmingham Railways Coach & Wagon Co Ltd
L&YR coach, Dinting Railway Centre

Lakeside & Haverthwaite Railway

Originally this Furness Railway branch line carried passengers and freight from Ulverston to Lakeside but now the only part remaining is the 3½-mile section from Haverthwaite to the terminus at Lakeside where connections are made with the 'Windermere Cruises' which ply the 10-mile length of Lake Windermere
Headquarters: Haverthwaite station, near Ulverston, Cumbria LA12 8AL
Telephone: Newby Bridge (05395) 31594
Main station: Haverthwaite
Other public stations: Intermediate station at Newby Bridge. Terminus at Lakeside
OS reference: SD 349843
Car parks: Haverthwaite, Lakeside
Access by public transport: Windermere Cruises ships on Lake Windermere call at Lakeside. Ribble bus to Haverthwaite
Refreshment facilities: Haverthwaite
Souvenir shop: Haverthwaite
Depot: All rolling stock at Haverthwaite
Length of line: 3½ miles
Passenger trains: Steam-hauled Haverthwaite-Lakeside
Period of public operation: Easter then daily from early-May to end of October.
Special events: Santa Specials (advance booking essential) please contact for details
Special notes: Combined railway/ lake steamer tickets available, from the station at Haverthwaite and

Locomotives

Name	*No*	*Origin*	*Class*	*Type*	*Built*
—	2073	LMS	4MT	2-6-4T	1950
—	42085	LMS	4MT	2-6-4T	1951
—	8(D2117)	BR	03	0-6-0DM	1959
—	17(AD601)	LMS	—	0-6-0DE	1945
—	03.072	BR	03	0-6-0DM	1959
—	5643	GWR	5600	0-6-2T	1925

Industrial locomotives

Name	*No*	*Builder*	*Type*	*Built*
*Caliban**	1	Peckett (1925)	0-4-0ST	1937
Rachel	9	M/Rail (2098)	4wDM	1924
Repulse	11	Hunslet (3698)	0-6-0ST	1950
Princess	14	Bagnall (2682)	0-6-0ST	1942
Askam Hall	15	Avonside (1772)	0-4-0ST	1935
Alexandra	12	Barclay (929)	0-4-0ST	1902
David	13	Barclay (2333)	0-4-0ST	1953
Cumbria	10	Hunslet (3794)	0-6-0ST	1953
—	7	Fowler (22919)	0-4-0DM	1940
Fluff	16	Hunslet/Fowler	0-4-0DM	1937
—	20	Jones crane	0-4-0DM	1952
—	21	Barclay (1550)	0-6-0F	1917
—	22	Fowler (4220045)	0-4-0DM	1967

*Under restoration at Steamtown, Carnforth

Stock
10 ex-BR Mk 1 coaches
2-car Class 110 DMU, vehicle Nos 52071, 52077
1 ex-LNER BG
1 ex-BR Mk 1 miniature buffet coach
Small selection of freight vehicles

'Windermere Cruises' piers at Bowness and Ambleside. 'Windermere Cruises' is operated by Windermere Iron Steamboat Co Ltd

Liverpool Museum M

A fine selection of vehicles in a traditional 'stuffed and mounted' display. *Lion* may be absent at certain times during the year, depending on developments. Visitors are advised to enquire by telephone as to her availability
Location: William Brown Street, Liverpool
Operating society/organisation: National Museums and Galleries on Merseyside, William Brown Street, Liverpool L3 8EN
Telephone: 051-207 0001
Car parks: Public car parks nearby
Access by public transport: Adjacent Lime Street stations (BR and Merseyrail). Numerous bus routes
On site facilities: Cafe, shop
Public opening: Please contact for details

Locomotives

Name	*No*	*Origin*	*Class*	*Type*	*Built*
*Lion**	57	L&MR	—	0-4-2	1838

**Lion* is the oldest working locomotive in the world

Industrial locomotives

Name	*No*	*Builder*	*Type*	*Built*
—	1	Avonside (1465)	0-6-0ST	1904

Stock
LOR electric coach of 1893
L&MR replica coach built Derby 1929

Special notes: Transport gallery within large general museum. Facilities for disabled visitors

Ravenglass & Eskdale Railway

Originally built to serve iron ore mines in Eskdale, this delightful line makes an ideal 'tourist' line running as it does through wooded valleys and along rugged hillsides
Headquarters: Ravenglass & Eskdale Railway, Ravenglass station, Cumbria
Telephone: 0229 717171
Main station: Ravenglass
Other public stations: Muncaster Mill, Irton Road, The Green, Eskdale (Dalegarth)
OS reference: SD 086964
Car parks: All stations
Access by public transport: BR services to Ravenglass
Refreshment facilities: Ravenglass, Dalegarth. Bar meals at 'Ratty Arms'
Souvenir shops: Ravenglass, Dalegarth
Museum: Ravenglass
Length of line: 7 miles, 15in gauge
Passenger trains: Steam or diesel-hauled narrow gauge trains Ravenglass-Dalegarth
Period of public operation: Daily from late March-late October. Limited winter service November-March
Special events: Family Day 25 May
Facilities for disabled: Special coaches for wheel-chair passengers. Advance notice preferred. Wheelchair access to toilets and museum at Ravenglass; toilets, shop and cafe at Eskdale (Dalegarth)
Special notes: At Ravenglass the R&ER has two camping coaches and the company also operates the 'Ratty Arms' public house formed by conversion of the former BR station buildings. During the high summer, mid-July through August, five steam locomotives are normally in use Monday-Thursday
Membership details: A. N. Grayston, 'Brunel', 10 Harriot Hill, Cockermouth, Cumbria CA13 0BL

Locomotives

Name	*No*	*Builder*	*Type*	*Built*
River Irt	—	Heywood	0-8-2	1894
River Esk	—	Davey Paxman (21104)	2-8-2	1923
River Mite	—	Clarkson (4669)	2-8-2	1966
Northern Rock	—	R&ER	2-6-2	1976
Bonny Dundee	—	K/Stuart (720)*	0-4-2WT	1901
Shelagh of Eskdale	—	R&ER/Severn-Lamb	4-6-4DH	1969
Quarryman	—	Muir-Hill (2)	0-4-0P/Paraffin	1928
Perkins	—	Muir Hill (NG39A)	0-4-4DM	1929
Silver Jubilee	—	R&ER	DMU	1977
Lady Wakefield	—	R&ER	Bo-Bo	1980
Synolda	—	Bassett-Lowke	4-4-2	1912
—	—	Greenbat (2782)	0-4-0B	1957
Cyril	—	Lister	0-4-0DM	1987

*Rebuilt to 15in gauge 1981

South Tynedale Railway

A narrow gauge line passing through the attractive scenery of the South Tyne valley, in the North Pennine area of outstanding natural beauty
Location: Approximately ¼-mile north of Alston town centre, on A686 Hexham road
OS reference: NY 717467
Operating society: South Tynedale Railway Preservation Society, The Railway Station, Alston, Cumbria CA9 3JB
Telephone: Alston (0434) 381696
Car park: Alston station
Access by public transport: Wright Bros buses, Haltwhistle-Alston and Newcastle-Alston-Keswick. *Enquiries:* Tel: Alston (0434) 381200. Also summer Sunday buses from Stanhope in connection with 'Heritage Line' trains. Details: (091) 386 4411 ext 2498. Bus links from Langwathby in connection with Settle-Carlisle line trains on certain dates (details: (0539) 721000, ext 4350)
On site facilities: Book and souvenir shop, tourist information centre, refreshments, picnic area, toilets (including disabled persons)
Catering facilities: Tea room at Alston, serving selection of snacks and home baking
Length of line: 1½ miles, 2ft gauge
Public opening: Passenger train service from Alston to Gilderdale: March-April — Easter week then Sundays; May — Saturday, Sundays, May Day Holiday, Spring Bank Holiday Week; June and September — Tuesdays, Wednesdays, Thursdays, Saturdays, Sundays; July and August — daily; October — Sundays. Departures from 11.15 until 16.00 or 17.00. Steam haulage scheduled at weekends from Spring Bank Holiday until end of September, also Easter and May Day weekends
Journey time: c40min (round trip)
Special events: Children's Day — 29 May; Steam Enthusiasts' Weekend — 27/28 July; Diesel Day — 13 October; Santa Specials — 8, 14/15, 22 December
Special notes: The line has been constructed on the trackbed of the former BR Haltwhistle-Alston branch. Extension of the line from Gilderdale towards Slaggyford, a further 3 miles, has commenced
Membership details: Membership Secretary c/o above address
Feature article reference: RW February 1989, p92

Locomotives

Name	*No*	*Builder*	*Type*	*Built*
Phoenix	1	Hibberd (2325)	4wDM	1941
Ayle	2	Hudson-Hunslet (2607)	4wDM	1942
Sao Domingos	3	O/K (11784)	0-6-0WT	1928
Naworth	4	H/Clarke (DM819)	0-6-0DM	1952
Thomas Edmondson	6	Henschel (16047)	0-4-0T	1918
—	9	Hunslet (4109)	0-4-0DM	1952
Naklo	10	Chrzanow (3459)	0-6-0WTT	1957
—	11	Hunslet (6646)	0-4-0DM	1967
Tiny Tim	—	Hunslet (6619)	0-4-0DM	1966

Locomotive notes: In addition to several of the above, the visiting Hunslet steam locomotive *Chaka's Kraal No 6* is expected to be in service during the 1991 season

Stock

7 bogie coaches
1 brake van
1 bogie box van
1 bogie open wagon
5 4-wheel open wagons
1 4-wheel box van
1 4-wheel flat wagon
1 fuel tank wagon
2 bogie well wagons
6 bogie hopper wagons

Owners

2, 3, 4, 9 and 11 (the Durham Narrow Gauge Group)
Tiny Tim (the Ayle Colliery Co)

Southport Railway Centre

SC

Exhibits include ex-Liverpool Riverside signalbox and the restored turntable facilities using the turntable from York (Queen St). A working ex-Midland Railway water column from London St Pancras is a dominant feature of the site. Steamport has been a servicing point for the annual Southport-Manchester steam specials since 1985

Locomotives

Name	*No*	*Origin*	*Class*	*Type*	*Built*
Cecil Raikes	5	Mersey Railway	—	0-6-4T	1886
—	5193	GWR	5101	2-6-2T	1934
—	D2148	BR	03	0-6-0DM	1960
—	D2953	BR	05	0-6-0DM	1959
—	24081	BR	24	Bo-Bo	1958

Location: Five minutes from BR station, via London Street. Behind car park
OS reference: SD 341170
Operating society/organisation: Southport Railway Centre, Derby Road, Southport PR9 0TY
Telephone: Southport (0704) 530693
Car park: Derby Road
Access by public transport: BR station, Merseyrail from Liverpool
On site facilities: Souvenir shop. Cafe open Sundays and Bank Holidays at weekends. Steam hauled rides in LMS brake vans June-end September, plus special events. Buffet car in operation on special events. Santa Steam trains in December
Length of line: 600yd
Public opening: October-April 13.00-17.00; Weekdays June-mid September 13.00-16.30 (July-August 10.30-16.30)
Facilities for disabled: Access to all parts of museum. Limited access to toilets. No advanced notice required for disabled visitors — groups always welcome. Enquiries always welcomed

Industrial locomotives

Name	*No*	*Builder*	*Type*	*Built*
—	5	Peckett (2153)	0-6-0ST	1954
North Western Gas Board	—	Peckett (1999)	0-4-0ST	1941
Fitzwilliam	—	Hunslet (1954)	0-6-0ST	1939
—	—	Hunslet (3155)	0-6-0ST	1943
Glasshoughton	4	Hunslet (3855)	0-6-0ST	1954
Persil	—	Fowler (4160001)	0-4-0DM	1952
*Efficient**	—	Barclay (1598)	0-4-0ST	1918
Agecroft No 2	—	RSH (7485)	0-4-0ST	1948
St Monans	—	Sentinel (9373)	4wVBT	1947
Hornet	—	Peckett (1935)	0-4-0ST	1937
Whitehead	—	Peckett (1163)	0-4-0ST	1908
—	—	Greenwood & Batley (2000)	0-4-0BE	1958
Sefton	—	T/Hill (123V)	4wDM	1963
Stanlow No 4	4	T/Hill (160V)	0-4-0DM	1966
—	—	N/British (27653)	0-4-0DH	1956
—	D2870	YEC (2677)	0-4-0DM	1960

Stock
Small selection of rolling stock including Class 502 EMU, M28361M and M29896M on loan from NRM, and buffet car Sc1839 and BSK 35128
1 Smith-Rodley diesel crane

Owner
Cecil Raikes (Liverpool Museum)
*Stock (Liverpool Locomotive Preservation Group)

Steamtown, Carnforth SC

Formerly BR Carnforth motive power depot. Carnforth is a centre for 'Steam on BR' steam-hauled railtours over BR main lines.
Location: Steamtown Railway Centre, Warton Road, Carnforth, Lancashire LA5 9HX
OS reference: SD 496708
Operating society/organisation: Steamtown Railway Museum Ltd
Telephone: Carnforth (0524) 732100
Car park: Free car park on site
Access by public transport: BR services to Carnforth station. Bus services to Carnforth by Ribble and Lancaster City Transport
Catering facilities: Buffet car for meals and snacks, several picnic areas
On site facilities: Model railway, signalbox, collectors' corner, shop, cafe, 15in gauge railway
Length of line: Standard gauge ¾-mile, 15in gauge 1-mile
Public opening: Weekends all year round, daily Easter-October. Summer 09.00-17.00, winter 10.00-16.00. Locomotives in steam on Bank Holidays, Sundays from Easter-October, and daily during July-August
Facilities for disabled: Limited, toilets available, standard gauge trains can accommodate wheelchairs
Special notes: Scheduled steam hauled excursions during summer season, operate from Carnforth with through connections from main centres
Feature article reference: RW September 1990, p525 (No 4498)

Locomotives

Name	*No*	*Origin*	*Class*	*Type*	*Built*
*Britannia**	70000	BR	7MT	4-6-2	1951
Flying Scotsman†	4472	LNER	A3	4-6-2	1923
Sir Nigel Gresley†	4498	LNER	A4	4-6-2	1937
Lord Nelson	850	SR	LN	4-6-0	1926
Sovereign	44871	LMS	5MT	4-6-0	1945
—	44932‡	LMS	5MT	4-6-0	1945
—	48151	LMS	8F	2-8-0	1942
—	5407	LMS	5MT	4-6-0	1937
—	6441	LMS	2MT	2-6-0	1950
—	1300	L&Y	27	0-6-0	1896
—	D2381	BR	03	0-6-0DM	1960
—	03120	BR	03	0-6-0DM	1959
—	08220	BR	08	0-6-0DE	1956
—	D5500	BR	31	A1A-A1A	1957
La France	231K22	SNCF	231K	4-6-2	1914
—	01.2104-6	DB	01	4-6-2	1940

*On site for restoration to BR standards for main line operation
†Sometimes based in London area for main line running
‡Shares time between here and Midland Railway Centre

Above:
Gresley 'A4' Pacific No 4498 *Sir Nigel Gresley* storms Hatton bank with the 'White Rose' rail tour after stopping at an adverse signal at the foot of the bank, photographed on a cold but sunny Saturday, 6 February 1988. Based at Steamtown the locomotive is currently undergoing a major overhaul. *Jonathan Webb*

Industrial locomotives

Name	*No*	*Builder*	*Type*	*Built*
Sirapite	6158	A/Porter (6158)	0-4-0TG	1906
Pitsford	—	Avonside (1917)	0-6-0ST	1923
John Howe	5	Barclay (1147)	0-4-0ST	1908
Coronation	3	Barclay (2134)	0-4-0ST	1942
Cooke & Nuttal	1	Barclay (2230)	0-4-0ST	1947
British Gypsum	4	Barclay (2343	0-4-0ST	1953
Glenfield	1	Barclay (880)	0-4-0CT	1902
May	—	Peckett (1370)	0-4-0ST	1915
Trimpell	—	Bagnall (3019)	0-6-0F	1953
Cranford No 2	—	Bagnall (2668)	0-6-0ST	1942
Great Central	39	RSH(6947)	0-6-0T	1938
Barrow Steel No 7	FR 18	S/Stewart (1435)	0-6-0ST	1863
Barrow Steel No 17	FR 25	S/Stewart (1585)	0-4-0ST	1865
Lindsay	—	Wigan Coal & Iron Co	0-6-0ST	1887
Joyce	—	Sentinel (7109)	4wVBTG	1950
Gasbag	—	Sentinel (8024)	4wVBTG	1923
Esso	—	General Electric (30483)	Bo-Bo	1949
Lord Trenchard	—	Barclay (401)	0-4-0DM	1956
Tom Rolt	7049	Hunslet (2697)	0-6-0DM	1944

Stock
4 pre-Grouping vehicles
1 GWR saloon
5 LNER coaches
2 LMS coaches
6 BR coaches
Small selection of freight vehicles
3 steam cranes
1 diesel crane

Owners
1300 (the Fairclough Corporation)
850 and D5500 (on loan from the National Railway Museum)

West Lancashire Light Railway — SC

Location: Alty's Brickworks. Station Road, Hesketh Bank, near Preston, Lancashire
OS reference: SD 448229
Operating society/organisation: The West Lancashire Light Railway Association, Secretary, 790 Ormskirk Road, Pemberton, Wigan WN5 8AX
Telephone: (0942) 218078 (Secretary)
Car parks: On site
Access by public transport: BR to Preston or Southport. Bus route 100 and 102. Between Preston and Southport
On site facilities: Gift shop, light refreshments, 2ft gauge line
Public opening: Easter weekend (not Saturday); May Day Bank Holiday Sunday and Monday; Spring Bank Holiday, then every Sunday until end of September. Opening times 13.00-17.30
Special events: Enthusiast's Day 14 July; Santa Specials 8, 15, 22 December

Industrial locomotives

Name	*No*	*Builder*	*Type*	*Built*
Clwyd	1	R/Hornsby (264251)	4wDM	1951
Tawd	2	R/Hornsby (222074)	4wDM	1943
Irish Mail	3	Hunslet (823)	0-4-0ST	1903
—	4	Hibberd (1777)	4wPM	1931
—	5	R/Hornsby (200478)	4wDM	1940
—	7	M/Rail (8992)	4wDM	1946
—	8	H/Hunslet (4480)	4wDM	1953
—	9	K/Stuart (2405)	0-6-0T	1915
—	10	Hibberd (2555)	4wDM	1946
—	16	R/Hornsby (202036)	4wDM	1941
Trent	18	R/Hornsby (283057)	4wDM	1949
—	19	Lister (10805)	4wPM	1939
—	20	Baguley (3002)	4wPM	1937
—	21	H/Hunslet (1963)	4wDM	1939
—	25	R/Hornsby (297054)	4wDM	1950
—	26	M/Rail (11223)	4wDM	1963
Mill Reef	27	M/Rail (7371)	4wDM	1939
Red Rum	29	M/Rail (7105)	4wDM	1936
—	30	M/Rail (11258)	4wDM	1964
Montalban	34	O&K (6641)	0-4-0WT	1913
Utrillas	35	O&K (2378)	0-4-0WT	1907
—	36	R/Hornsby (339105)	4wDM	1953
—	38	Hudswell (D750)	0-4-0DM	1949
*Jonathan**	—	Hunslet (678)	0-4-0ST	1898

*On loan from The Lytham Creek Motive Power Museum

Note: 30 M/Rail (11258) is presently on loan to Steamport Southport

Stock
Toastrack coach built 1986 by WLLR
Brake van built 1987 by WLLR
Large collection various wagons

Below:
***Irish Mail*, yet another Hunslet-built 0-4-0ST, in action on the West Lancashire Light Railway.**

North East

Beamish SC

The railway station, signalbox and goods shed have been completely recreated along with the other exhibits to show a way of life long past. There are some very old locomotives in the collection

Location: The North of England Open Air Museum, Beamish, County Durham DH9 0RG

OS reference: NZ 217547

Telephone: Stanley (0207) 231811

Car park: At museum

Access by public transport: Bus service from Worswick Street, Newcastle upon Tyne; Bus service Nos 775 and 778 from Sunderland via Chester-le-Street; Bus service 720 from Milburngate, Durham City

On site facilities: 200-acre open air museum vividly recreates life in the North of England around the turn of the century. The town with dentist's home and surgery, solicitor's office, the Sun Inn public house, Co-operative shops and printer's workshop. The railway station with country station, goods yard, signalbox and rolling stock and locomotives often in steam. The Colliery Village with furnished pit cottages, 'drift' mine and colliery buildings. Home Farm with farm house, livestock and exhibitions. Tea Room, souvenir shop and picnic areas

Public opening: April-October (inclusive) daily 10.00-18.00. November-March (inclusive) daily except Mondays 10.00-17.00. Last admission is always 4pm. Please check for Christmas opening times. Locomotives do not steam, and some areas are closed in winter months

Special events: A full programme of events is planned for 1991

Length of line: Approx ½-mile single line connecting rebuilt NER station to colliery sidings. No through passenger trains

Facilities for disabled: Not ideal for wheelchairs. Rolling stock not converted. Advanced notice for parties to Bookings Officer preferred

Locomotives

Name	*No*	*Origin*	*Class*	*Type*	*Built*
—	876	NER	C1	0-6-0	1889
Locomotion	1*	—	—	0-4-0	1975

*Replica, and may be out on long-term loan

Industrial locomotives

Name	*No*	*Builder*	*Type*	*Built*
Twizell	3	Stephenson (2730)	0-6-0T	1891
—	14	H/Leslie (3056)	0-4-0ST	1914
*South Durham**				
Malleable	No 5	Stockton Ironworks	0-4-0ST	c1880
Coffee Pot	—	Head Wrightson	0-4-0VB	1871
—*	E1	Black, Hawthorn (897)	2-4-0CT	1883
Hetton Loco	—	G. Stephenson (1)	0-4-0	1822
—	—	R/Hornsby (476140)	0-4-0DM	1963
Jacob	680	McEwan Pratt	0-4-0P	1916

*In store for long-term restoration

Owner

Hetton Loco (on loan from National Railway Museum)
Locomotion (the Locomotion Trust)

Note

Not all exhibits on display

Special notes: 1991 'Season of Celebration' for the 21st Birthday of Beamish

Bowes Railway SC

The railway includes the only preserved rope-hauled standard gauge inclines, whose operation requires considerable skill and sleight of hand. You should not miss the opportunity of inspecting the inclines and cable house and haulage engine when you can

Location: Bowes Railway, Springwell Village, near Gateshead, Tyne & Wear (on B1288)
OS reference: NZ 285589
Operating society/organisation: Bowes Railway Co Ltd
Telephone: Tyneside (091) 4161847 (Answerphone)
Car park: Springwell
Access by public transport: Northern Buses services Nos 184 Washington/Birtley, 187/188 Gateshead Metro/Sunderland, 189 Washington (Brady Sq)-Gateshead 638 Ryton/Sunderland
On site facilities: Exhibition of Railway's history, wagon exhibition, workshop displays. On operating days — shop, refreshments and guided tours. Steam hauled brake van rides. Rope haulage demonstration trains
Public opening: Bank Holidays, also first and third Sundays in each month, Easter-September (tel: (091) 2622651 for confirmation)
Length of line: 1¼ miles total length; 1¼ miles of rope haulage; ¾-mile used for passenger trains
Special notes: Preserved section of the Pontop & Jarrow Railway; designed G. Stephenson; opened 1826; the only standard gauge rope-hauled incline railway in the world; Railway's own historic workshops preserved, with examples of all of the Railway's wagon types
Facilities for disabled: Toilet only

Industrial locomotives

Name	*No*	*Builder*	*Type*	*Built*
WST	—	Barclay (2361)	0-4-0ST	1954
—	22	Barclay (2274)	0-4-0ST	1949
—	20/110/709	Barclay (613)	0-6-0DH	1977
—	6263	Hunslet	0-4-0DM	1964
Norwood	77	RSH (7412)	0-6-0ST	1948
—	101	Planet (3922)	4wDM	1959
—	—	Smith (18773)	4wCT	1948

Owners
22 (On loan from Thomas Ness Ltd)
WST (On loan from British Gypsum Ltd)
Norwood (On loan from National Smokeless Fuels)
Barclay 0-6-0DH (On loan from Mr P. Dawe)
Hunslet (On loan from Coal Products, Ness)
Planet and Smith (Bowes Railway Co Ltd)

Stock
20 Ordinary 10-ton wooden hopper wagons (Springwell built)
16 other wooden hopper wagons (of various pedigrees)
3 steel 14-ton hopper wagons
2 steel 21-ton hopper wagons
1 reel bogie (for rope replacement)
1 drift bogie (for shunting by rope)
1 loco coal wagon
7 material wagons
1 tool van
3 brake vans
2 flat wagons
1 17-ton wooden hopper (ex-NER)
1 18-ton wooden hopper (ex-Ashington)
1 21-ton wooden hopper (ex-Seaham)

Stationary haulage
Met-Vick/Wild, 300bhp electric (Blackham's Hill) 1950
BTH/Robey, 500bhp electric (Black Fell) 1950
Clarke Chapman, 22hp electric (Springwell Yard)
14ft Diam, Gravity Dilly Wheel (Springwell)

Membership details: Management Committee Secretary, c/o above address or telephone number
Disclaimer: The Bowes Railway Co Ltd wish to point out that all advertised facilities are subject to alteration without prior notice. The company can therefore not be held responsible for any loss or expense incurred

Darlington Railway Centre and Museum M

Located on the original 1825 route of the Stockton & Darlington Railway, the restored North Road station, dating from 1842, is now a museum which forms the centrepiece of an area devoted to railway history and preservation. The collection includes locomotives, rolling stock and many small exhibits. The site also includes

Locomotives

Name	*No*	*Origin*	*Class*	*Type*	*Built*
Locomotion	1	S&DR	—	0-4-0	1825
Derwent	25	S&DR	—	0-6-0	1845
—	1463	NER	1463	2-4-0	1885
—	910	NER	901	2-4-0	1875

two other historic buildings of the S&DR — the former Goods Shed and the Hopetown Carriage Works. Steam train rides over a short length of line are available on selected dates, and there are plans for future extension of the line.

The Ken Hoole Study Centre, housing a collection of reference material, will open during 1991 (access by appointment). British Rail's 'Heritage Line' provides a link to Darlington's main line station and to Shildon, for the Timothy Hackworth Museum

Location: North Road station, Darlington, County Durham DL3 6ST. Approximately ¾-mile north of town centre, off North Road (A167)
OS reference: NZ 289157
Telephone: 0325 460532
Car park: At museum site
Access by public transport: BR services to Darlington North Road station. Local bus services along North Road
Catering facilities: Confectionery and drinks. Party catering by arrangement
On site facilities: Souvenir and book shop. Toilets
Public opening: Daily throughout the year 09.30-17.00 (except Christmas/ New Year holidays). Last admission 16.30 (times may be subject to amendment)
Special events: Steam days and Railway Carnival (contact for details)
Facilities for disabled: Access to main museum building for wheelchairs. Disabled persons toilet
Membership details: Friends of Darlington Railway Museum and Darlington Railway Preservation Society, both c/o above address
Note: Some locomotives are located in the former goods shed, where restoration work is being undertaken by the Darlington Railway Preservation Society (limited opening to visitors — times vary; groups by arrangement)

Industrial locomotives

Name	*No*	*Builder*	*Type*	*Built*
Met	—	H/Leslie (2800)	0-4-0ST	1909
—	17	Head Wrightson (33)	0-4-0VB	1873
—	—	Bagnall (2898)	0-4-0F	1948
—	39	RSH (6947)	0-6-0T	1938

Stock
1 Stockton & Darlington Rly passenger coach (1846)
1 North Eastern Railway Coach body (c1860)
1 NER 20-ton mineral wagon
1 Chaldron wagon

Owners
Locomotion, Derwent, 1463 and 910 are all on loan from the National Railway Museum
Met is on loan from Messrs D. & R. Branch

Darlington Railway Preservation Society
Locomotives

Name	*No*	*Origin*	*Class*	*Type*	*Built*
—	78018	BR	2MT	2-6-0	1954

Industrial locomotives

Name	*No*	*Builder*	*Type*	*Built*
—	2	RSH (7925)	0-4-0DM	1959
—	1	Peckett (2142)	0-4-0ST	1953
David Payne	185	Fowler (4110006)	0-4-0DM	1950
Smiths Dock Co Ltd	—	Fowler (4200018)	0-4-0DM	1947
—	—	GEC	4wE	1928
—	—	R/Hornsby (279591)	0-4-0DM	1949

Stock
Various wagons, steam and diesel cranes

Embsay Steam Railway SC

Yorkshire's 'Friendly Line' is operated by volunteers whose whole aim is to reopen the line through to Bolton Abbey. A half-mile extension of the line towards Bolton Abbey opens for the 1991 season. An atmosphere of the days of the rural branch line prevails
Location: Embsay station, Embsay, Skipton, Yorkshire BD23 6QX
OS reference: SE 007533
Operating society/organisation: Yorkshire Dales Railway Museum Trust
Telephone: Skipton (0756) 4727. 24hr Answerphone (0756) 5189
Car park: Embsay
Access by public transport: Pennine bus from Skipton

Locomotives

Name	*No*	*Origin*	*Class*	*Type*	*Built*
—	D2203	BR	04	0-6-0DM	1952
—	NCB 38 (D9513)	BR	14	0-6-0DH	1964

Industrial locomotives

Name	*No*	*Builder*	*Type*	*Built*
Annie	—	Peckett (1159)	0-4-0ST	1908
Gladiator	—	H/Clarke (1450)	0-6-0ST	1922
Slough Estates No 5	—	H/Clarke (1709)	0-6-0ST	1939
Primrose No 2	—	Hunslet (3715)	0-6-0ST	1952
Ann	—	Sentinel (7232)	4wVB	1927
Beatrice	S119	Hunslet (2705)	0-6-0ST	1945
Airedale	—	Hunslet (1440)	0-6-0ST	1923
City Link	—	Yorkshire (2474)	0-4-0ST	1949

On site facilities: Souvenir shop — specialising in transport and industrial archaeological titles, plus model railway supplies, small museum and lead mining centre
Catering facilities: Buffet and real ale bar on most trains. Buffet on station. Evening catering trains will serve meals on selected dates. Special charters can be arranged. Meals for parties can be arranged on normal service trains (subject to advance booking). Please write for further details
Length of line: 2½ miles
Public opening: Trains will run, Sundays throughout the year, plus Saturdays in July and August. Tuesdays in July and August. Wednesdays and Thursdays August only. Also all Bank Holiday weekends. Site open all year until dusk. Train times at least hourly 11.00-16.15
Special events: 31 March, 1 April — Easter Egg Specials; 6 May, 16 June — Kiddies' Day; 14 July — Steam Gala; 28 July, 18 August — Kiddies' Day; 8 September — Teddy Bear's Picnic; 27 October — Halloween Specials; 2 November — Bonfire Night; 17, 24 November, 1, 8, 15, 22 December — Santa Trains; 1 January 1992 — Family Day
Special notes: Steam rides are on 2-mile line to new halt and picnic area. Unlimited ride facility. Old Midland Rly buildings, fine collection of industrial locomotives
Membership details: Membership Secretary at above address

—	140	H/Clarke (1821)	0-6-0T	1948
Spitfire	S112	Hunslet (2414)	0-6-0ST	1942
Wheldale	—	Hunslet (3168)	0-6-0T	1944
—	69	Hunslet (3785)	0-6-0ST	1953
Monkton No 1	—	Hunslet (3788)	0-6-0ST	1953
—	22	Barclay (2320)	0-4-0ST	1952
Thomas	4	RSH (7661)	0-4-0ST	1950
H. W. Robinson	—	Fowler (4100003)	0-4-0DM	1946
—	MDE15	Baguley/Drewry (2136)	4wDM	1938
—	887	R/Hornsby (394009)	4wDM	1955
—	—	Wickham (7610)	2w-2PMR	1957
—	—	Lister (9993)*	4wPM	1938
—	—	Lister (10225)*	4wPM	1938
—	—	R/Hornsby (175418)*	4wDM	1936
—	—	R/Hornsby*	4wDM	—
—	—	M/Rail (8979)*	4wDM	1946
—	—	M/Rail (5213)*	4wDM	1930
—	—	R/Hornsby	4wDM	1957

*2ft gauge

Stock
11 ex-BR Mk 1 coaches (SK, CK, BCK, 5×TSO, RMB, BSO(T) and SLS)
4 ex-LNER coaches
2 SR parcels vans
Freight stock and service vehicles, SR and GW brakes

Right:
Two-train working on 12 August 1990, a 'Thomas' event, witnesses *Beatrice* running into Platform 2 at Embsay. The shed and stock in the background are just two requirements to run a preserved line. *Melvyn Hopwood*

Keighley & Worth Valley Railway

The epitome of a volunteer-run railway, a thriving branch line which serves six stations (most of them gas lit) which is host to an extensive and varied collection of locomotives and where everything continues to provide the atmosphere of the days of the steam railway. Immaculate stations, immaculate staff, immaculate trains. Very friendly and very much part of the community

Headquarters: Haworth station, Keighley, West Yorkshire, BD22 8NJ

Telephone: Haworth (0535) 643629 24hr recorded timetable and information service, Haworth (0535) 645214 (other calls)

Main stations: Keighley, Ingrow West, Haworth, Oxenhope

Other public stations: Damems, Oakworth

OS reference: SE 034371

Car parks: Free at Keighley, Ingrow West, Oakworth, Oxenhope. Limited parking at Haworth (pay). *Coaches* at Ingrow West and Oxenhope only

Access by public transport: British Rail through services from Carlisle, Morecambe, Lancaster, Leeds, Bradford to Keighley station for Worth Valley trains. (Change to platform 4). Also from Preston, Blackburn, Accrington, Burnley, Manchester, Rochdale, Todmorden to Hebden Bridge Station for connection via bus service 500 to Oxenhope. (Operates Saturdays throughout the year, Sundays May to September, Monday to Friday July and August only. Tel: Wakefield (0924) 375555 for timings.) A large number of bus services serve Keighley including from Burnley; Burnley & Pendle/Keighley & District Service 25 to Keighley bus station. From Bradford Interchange: services 663/664/665/666/667/668/669/699. From Leeds: service 760. From Huddersfield and Halifax: service 502 via Denholme to Ingrow West. National Coach services Nos 281/2/5/9, 291, 307/8, 318 all serve Keighley from various parts of the country

Refreshment facilities: Buffet restaurant at Oxenhope. Buffet at Keighley station. All normally open when steam trains are operating. Buffet bar car on some trains. Wine and Dine by prior booking only — the 'White Rose Pullman'; Sunday lunch dining car services on the 'Devonian'

Souvenir shops: Haworth, Oxenhope, Keighley, Ingrow carriage museum

Locomotives

Name	*No*	*Origin*	*Class*	*Type*	*Built*
—	41241	LMS	2MT	2-6-2T	1949
—	43924	MR	4F	0-6-0	1920
—	45212	LMS	5MT	4-6-0	1935
Bahamas	45596	LMS	Jubilee	4-6-0	1935
—	8431	LMS	8F	2-8-0	1944
—	47279	LMS	3F	0-6-0T	1925
—	1054	LNWR	—	0-6-2T	1888
City of Wells	34092	SR	WC	4-6-2	1949
—	80002	BR	4MT	2-6-4T	1952
—	75078	BR	4MT	4-6-0	1956
—	78022	BR	2MT	2-6-0	1953
—	30072	SR	USA	0-6-0T	1943
—	5775	GWR	5700	0-6-0PT	1929
—	52044	L&Y	2F	0-6-0	1887
—	19	L&Y	Pug	0-4-0ST	1910
—	51218	L&Y	Pug	0-4-0ST	1901
—	752	L&Y	—	0-6-0ST	1881
—	85	TVR	02	0-6-2T	1899
—	5820	USA TC	S160	2-8-0	1945
—	90733	MoS	WD	2-8-0	1945
—	68077	LNER	J94	0-6-0ST	1947
—	D226	BR	—	0-6-0DE	1956
—	D2511	BR	—	0-6-0DM	1961
—	D3336	BR	08	0-6-0DE	1954
—	D5209	BR	25/1	Bo-Bo	1963

Industrial locomotives

Name	*No*	*Builder*	*Type*	*Built*
Hamburg	31	H/Clarke (697)	0-6-0T	1903
—	67	H/Clarke (1369)	0-6-0T	1919
*Nunlow**	—	H/Clarke (1704)	0-6-0T	1938
Brussels	118	H/Clarke (1782)	0-6-0ST	1945
Southwick	—	RSH (7069)	0-4-0CT	1942
Fred	—	RSH (7289)	0-6-0ST	1945
—	57	RSH (7668)	0-6-0ST	1950
—	63	RSH (7761)	0-6-0ST	1954
—	2226	Barclay (2226)	0-4-0ST	1946
Tiny	—	Barclay (2258)	0-4-0ST	1949
Merlin	231	H/Clarke (D761)	0-6-0DM	1951
Austins No1	—	Peckett (5003)	0-4-0DM	1961
—	MDHB No 32	Hunslet (2699)	0-6-0DM	1944

*On loan to Swanage Railway

Stock

2 ex-BR German-built Waggon und Maschinenbau diesel railbuses
30 coaches including examples of pre-Grouping types; BR Mk 1 stock including the oldest vehicle in existence, part of the prototype batch; a Pullman car, NER and L&Y observation cars

Owners

19, 752 and 51218 (L&Y Saddletanks)
47279 (South Yorkshire 3F Fund)
75078 and 78022 (the Standard 4 Preservation Society)
Bahamas, Nunlow, Tiny (the Bahamas Locomotive Society)
1054 (the National Trust)

Museums: Railway Museum at Oxenhope, Vintage Carriage Trust's carriage and locomotive museum at Ingrow. Open daily May-October, 12.00-16.00. Weekends throughout the year
Depots: Carriage and wagon — Oxenhope; Motive power/loco works — Haworth, 'Bahamas Locomotive Society' workshops at Ingrow West (open to the public)
Length of line: 4¾ miles
Passenger trains: Steam-hauled services between Oxenhope-Keighley. Some off-peak and local shoppers' services worked by diesel railbus

Free bus service between Haworth station and Haworth village top on Sundays (May-September) and bank holidays, 11.00-18.00
Period of public operation: Winter diesel railbus service Saturdays November-February (Keighley-Oxenhope). Winter steam service Sundays November, January and February (Keighley-Oxenhope); December — Santa Specials only from Haworth and Oxenhope only (pre-booked). Steam/diesel service Saturdays and Sundays March-October. Daily steam service throughout July and August, Easter week and Spring Bank Holiday week, Wednesdays in June
Special events: 2nd weekend before Easter; 'Complete Railway Weekend', railway horses, vintage bus services, Griddle and Restaurant car services, gas and oil lit stations, intensive steam services, working timetable, 08.00-19.00.

December weekends before Christmas, Santa Special services
Facilities for disabled: Toilet on Haworth station forecourt. Wheelchairs accommodated in guard's compartments on trains. Please advise before visit to the Advanced Bookings Manager c/o Haworth station

Above:
Nos 47279 and 5775 double-head the 12.13 Keighley-Oxenhope service on 1 April 1990. They are seen here approaching Mytholmes viaduct during K&WVR's enthusiast weekend. No 5775 sports the livery carried for the filming of *The Railway Children* in 1970.
Doug Birmingham

Special notes: Accompanied children under 5 years of age free. Children 5-15 and Senior Citizens at 50% discount. BR Family Railcard holders 10% discount on full-line tickets. Through bookings from all major BR stations to Haworth and Oxenhope, discount on KWVR portion of ticket
Membership details: Membership Secretary c/o above address
Feature article references: RW September 1988, p458 (locomotive 43924); RW February 1989, p78

Leeds Industrial Museum M

Location: The Leeds Industrial Museum, Armley Mills, Canal Road, Leeds LS12 2QF
OS reference: SE 275342
Operating society/organisation: Leeds City Council, Department of Leisure Services, 19 Wellington Street, Leeds 1
Telephone: (0532) 637861
Car park: Cark park adjacent to the Museum
Access by public transport: Nos 5 or 5A from City Square, Leeds (outside the railway station)
Public opening: April-September: Tuesdays-Saturdays 10.00-18.00, Sundays 14.00-18.00. October-March: Tuesdays-Saturdays 10.00-17.00, Sundays 14.00-17.00. Closed Mondays (except Bank Holidays)
On site facilities: Museum shop, refreshments (vending machines)
Special notes: Facilities for the disabled (toilets, etc), lifts. Museum can be viewed by visitors in wheelchairs (most areas are accessible)
Details of locomotive and rolling stock: Locomotive collection includes steam, diesel, mines locomotives and a narrow gauge railway and engines

Middleton Railway SC

This is a preserved section of one of the world's oldest railways, authorised by Act of Parliament in 1758, and also the first to be re-opened by volunteers
Headquarters: Middleton Railway Trust Ltd, Moor Road, Leeds LS10 2JQ
Telephone: (0532) 710320 (Ansaphone) or (0532) 711089 after 6pm
Main station: Moor Road, Hunslet
OS reference: SE 302309
Car parks: Tunstall Road/Moor Road (free)
Access by public transport: Bus No 61 from Park Row to MRT (every 10min); Bus 59 from Corn Exchange to Moor Road, ½-hourly on Saturday or hourly on Sundays; Bus 76 from Park Row (opposite City station) to Station Hotel, Hilledge Road. From bus stop, walk past Station pub, and across railway bridge. Follow footpath alongside dual carriageway and under roundabout. Moor Road station is just in front of you. Bus details correct at time of writing but due to changes taking place every few months, passengers are advised to check details on arrival in Leeds
Refreshment facilities: Moor Road, tea, coffee, snacks. Also café in Middleton Park (300yd)
Souvenir shop: Moor Road
Museum: Depot now open 10.00 to 18.00 during season (10.00 to 16.00 during winter). There is no charge for admission to site
Length of line: 1¼ miles
Passenger trains: Sunday trains all steam operated. Saturday services may be diesel hauled. Special trains operate on request, contact J. Wickinson, Tel: (0532)719785
Period of public operation: Every Saturday, Sunday and Bank Holiday Monday from 6 April to end of September. Normal Sunday steam timetable; 13.00 then ½-hourly until 16.30. Normal Saturday diesel timetable; 13.30 then every 45min until 16.30. Wednesdays 7, 14, 23, 28 August, steam hauled, 13.30 then every 45min until 16.30
Special events: 29 March-1 April — Easter train service; 4-6 May — May Day holiday trains; 15 May — Schoolday trains; 25-27 May — Spring Bank Holiday trains; 18/19 June — Schoolday trains; 30 June — 'Friends of Thomas' day; 4 August — Tank engine day; 24-26 August — bank holiday trains; 22 September — Gala Day; 1, 7/8, 14/15, 21/22 December — Santa trains
Facilities for disabled: Access very good
Special notes: It was the first standard gauge railway to be opened by volunteers, in June 1960. The railway terminates in a large car park and a nature trail has been provided by the local council
Feature article reference: RW May 1989, p268; RW August 1990, p468 (Then and Now — photo feature)

Locomotives

Name	*No*	*Origin*	*Class*	*Type*	*Built*
—	1310	NER	Y7	0-4-0T	1891
—	54	LNER	Y1	0-4-0VB	1933
—	385	DSB	HsII	0-4-0WT	1893
—	7401	LMS	—	0-6-0DM	1932

Industrial locomotives

Name	*No*	*Builder*	*Type*	*Built*
John Blenkinsop	—	Peckett (2003)	0-4-0ST	1941
Henry de Lacy II	—	H/Clarke (1309)	0-4-0ST	1917
Mirvale	—	H/Clarke (1882)	0-4-0ST	1955
Windle	—	Borrows (53)	0-4-0WT	1909
Matthew Murray	—	Bagnall (2702)	0-4-0ST	1943
Harry	—	A/Barclay (1823)	0-4-0ST	1926
Arthur	—	M/Wardle (1601)	0-6-0ST	1901
—	—*	Brush (91)	0-4-0DE	1958
—	No 6	H/Leslie (3860)	0-4-0ST	1935
Carroll	—	H/Clarke (D631)	0-4-0DM	1946
—	—	Hunslet (1786)	4wD	1935
Mary	—	H/Clarke (D577)	0-4-0DM	1932
—	—	Fowler (3900002)	0-4-0DM	1945
—	—	Fowler (4200038)	0-4-0DH	1966
—	—	Thomas Hill (138C)	0-4-0DH	1963
Rowntrees No 3†	—	R/Hornsby (441934)	4wDM	1960

*On loan from BSC Orb Works, Newport
†On loan from North Yorkshire Moors Railway

Stock
2 CCTs converted for passenger use Nos 1867 and 2048
CCT as stores van No 2073
Norwegian brake coach
Various goods vehicles
5-ton Booth rail crane
1 3-ton Smith steam crane
1 3-ton Isles steam crane

Owners
1310, 385 (the Steam Power Trust)
Harry (owned by Crossley Bros, Shipley)

Museum of Army Transport M

Operating a substantial network of railways in wartime the Army is still responsible for military trains in Germany and railways feeding MoD depots in this country. Not only does the Museum hold extensive archives and display some fascinating maps, drawings and photographs, it contains some very interesting stock from the former military railways
Location: Museum of Army Transport, Flemingate, Beverley
OS reference: TA 041392
Operated by: The Museum of Army Transport Ltd (Charitable Status Company)
Telephone: Hull (0482) 860445
Car park: Yes, 1¼ acres
Access by public transport: Rail: 10min walk from Beverley station. Bus: served by East Yorks Motor Service
On site facilities: Museum — The Royal Corps of Transport Collection of road, rail, sea, air and movement control artefacts. Licensed cafeteria and Beverley Bar. Shop in building. Lavatories, including special for handicapped
Facilities for disabled: The Museum is on one floor. This allows viewing of all vehicles. Access is available by ramp into the Armoured Train. Access is not possible into the Beverley aircraft, the railway locos and the signalbox frame
Public opening: 10.00-17.00 every day except 24-26 December

Locomotives

Name	*No*	*Builder*	*Type*	*Built*
Gazelle	—	Dodman	0-4-2WT	1893
Woolmer	—	Avonside (1572)	0-6-0ST	1910
Waggoner	92	Hunslet (3792)	0-6-0ST	1953
Rorke's Drift	—	Drewry (2047)	0-4-0DM	1934
—	110	R/Hornsby (411319)	4wDM	1958
—	1035	Wickham Rail Car	4w	1958
—	3282*	Wickham Target Trolley	4w	1943
—	WD2182†	M/Rail (461)	4wDM	1917
—	—†	M/Rail (3849)	4wDM	1927
—	LOD 758228†	M/Rail (8667)	4wDM	1941
—	LOD 758208†	M/Rail (8855)	4wDM	1943

*2ft 6in gauge
†2ft gauge

Rolling stock
Collection of various rolling stock items including Lord Kitchener's coach (c1885), a World War 1 Armoured Train gun truck and a World War 2 ramp wagon

Owners
Gazelle (the National Railway Museum)

A 2ft gauge system based on the ADLR of World War 1, some 200yd long, and worked by World War 2 Simplex rail tractors operated by volunteers, on Saturday and Sunday afternoons and at other times as advertised locally during the summer. Intending travellers should check the line is operating.

Special notes: Please bear in mind that while what is reported here deals with railways, other parts of this extensive museum cover movement by road, sea and air

National Railway Museum M

Location: National Railway Museum, Leeman Road, York YO2 4XJ
OS reference: SE 594519
Operating society/organisation: Part of the National Museum of Science and Industry
Telephone: York 621261 (STD Code 0904)
Car park: Limited parking for cars and pre-booked coaches at the Museum. York City's Leeman Road car park for cars and coaches visiting the NRM adjoins the main entrance
Access by public transport: The Museum is within a few minutes' walking distance of the railway and bus stations in York. The York City and District Bus Service operates to the door
On site facilities: Museum shop, restaurant and toilets, reference Library (free by appointment)
Public opening: Daily, weekdays 10.00-18.00. Sundays 11.00-18.00. Closed on New Years' Day, Christmas Eve, Christmas Day and Boxing Day
Facilities for disabled: All areas of the museum are accessible by the disabled
Special notes: The Museum has been open since 1975 and has welcomed over 15 million visitors.

A completely new and thematic exhibition, The Great Railway Show, opened on 1 March 1990 in the former York Goods Depot on the Museum's extensive 16-acre site. During the currency of the Show the Museum's Main Hall display area will be re-roofed.

The Great Railway Show illustrates the concept of travel by train — for passengers and freight. Short but representative trains will be drawn up at platforms and some locomotive footplates and carriages will be open to visitors. Outside in the yard there will be a regular programme of working locomotives — together with rides and footplate visits. A temporary display building will house a rotating exhibition of famous locomotives under the general title of 'Record Breakers and Workhorses'. Finally, the 'Magician's Road', an interactive educational exhibition, will provide booked school parties, and other visitors out of school hours, with 'hands-on' experience of various aspects of railway operation. Table 1 lists the locomotives expected to be on display during the Great Railway Show. A second major exhibition of National Railway Collection items was held at Swindon during 1990. The exhibits have returned to York, but are listed in Table 3. Inevitably, however, some items will remain in store and, for completeness, they are listed in Table 2. It must be emphasised that the appearance of any particular item on public display cannot be guaranteed. If it is vital to discover the exact whereabouts of a specific exhibit enquirers should contact the Museum in York *before* making a visit.

Table 4 is a list of National Railway Collection items on loan to other organisations. The list was correct in October 1989 but some changes may have occurred since.

The Museum is now open for evening-hire for private viewings and celebrations. the Great Railway Show offers unique opportunities for corporate hire; menu and details are available on request.

Feature article reference: RW September 1989 pp524-527 (The NRM's Development Project).

Membership details: Friends of the National Railway Museum c/o above address

TABLE 1

Exhibits at Great Railway Show, York

Locomotives

Name	*No*	*Origin*	*Builder*	*Class*	*Type*	*Built*
—	1	GNR	Doncaster	—	4-2-2	1870*
Boxhill	82	LB&SCR	Brighton	A1	0-6-0T	1880
Hardwicke	790	LNWR	Crewe	—	2-4-0	1892
—	1621	NER	Gateshead	M	4-4-0	1893
—	245	LSWR	Nine Elms	M7	0-4-4T	1897
—	673	MR	Derby	—	4-2-2	1899
—	1247	GNR	S/Stewart (4492)	J52	0-6-0ST	1899
—	737	SECR	Ashford	D	4-4-0	1901
—	1000	MR	Derby	4	4-4-0	1902
—	2818	GWR	Swindon	2800	2-8-0	1905*
—	1217	GER	Stratford	J17	0-6-0	1905
Cheltenham	925	SR	Eastleigh	V	4-4-0	1934
Rocket (replica)	—	Sectioned	R. Stephenson	—	0-2-2	1935
Green Arrow	4771	LNER	Doncaster	V2	2-6-2	1936*
Duchess of Hamilton	46229	LMS	Crewe	'Coronation'	4-6-2	1938*†
Mallard	4468	LNER	Doncaster	A4	4-6-2	1938*†
Eustace Forth	—	CEGB	RSH (7063)	—	0-4-0ST	1942
—	D8000	BR	E/Electric	20	Bo-Bo	1957
—	D200	BR	E/Electric	40	1Co-Co1	1958*
—	E5001	BR	Doncaster	71	Bo-Bo	1958
—	84001	BR	N/British	84	Bo-Bo	1960
The King's Own Yorkshire Light Infantry	55002	BR	E/Electric	55	Co-Co	1961*
Rocket (replica)	—	—	Loco Enterprises	—	0-2-2	1979
Iron Duke (broad gauge replica)	—	GWR	RESCO	—	4-2-2	1985

*To be displayed in rotation
†Will visit NRM on Tour exhibition, Swindon during 1990

Rolling Stock

Carriages

1869	LNWR Queen Victoria's Saloon
1885	MR 6-wheel composite brake No 901
1885	WCJS 8-wheel TPO No 186
1887	GNR 6-wheel brake No 948
1898	ECJS 3rd class No 12
1902	LNWR King Edward's saloon No 800
1902	LNWR Queen Alexandra's saloon No 801
1903	LSWR Tricomposite brake No 3598
1905	LNWR Royal service brake No 5154
1908	ECJS Royal Saloon No 395
1913	Pullman Car Co 1st class parlour car *Topaz*
1914	MR Dining car No 3463
1930	L&MR 1st class and 3rd class replicas
1936	CIWL Night Ferry sleeping car No 3792
1937	LNER Buffet Car No 9135
1985	GWR broad gauge carriage replica

Departmental stock

1906	NER Dynamometer car No 902502
1907	NER steam breakdown crane No CME 13
1949	BR (GWR design) inspection saloon No 80790

TABLE 2

In store at the National Railway Museum, York

Locomotives

Name	*No*	*Origin*	*Builder*	*Class*	*Type*	*Built*
—	3	FR	Bury Curtis & Kennedy	—	0-4-0	1846
Cornwall	3020	LNWR	Crewe	—	2-2-2	1847
—	84	—	B/Peacock	—	0-4-0 Tram	1886
Wren	—	LYR	B/Peacock	—	0-4-0ST	1887
Henry Oakley	990	GNR	Doncaster	C2	4-4-2	1898
—	75S	W&CR	Siemens (6)	—	Bo electric	1898
—	87	GER	Stratford	J69	0-6-0T	1904
—	102	GCR	Gorton	O4	2-8-0	1911
—	—	WD	Simplex (60cm gauge)	—	0-4-0	1918
—	—	Berry Wiggins & Co	Simplex	—	0-4-0	1931
—	KF7	Chinese Govt Rlys	Vulcan (4674)	KF	4-8-4	1935
—	5000	LMS	Crewe	5MT	4-6-0	1935
—	—	Yorks Water Auth	R/Hornsby (2ft gauge)	—	0-4-0	1937
Winston Churchill	34051	SR	Brighton	BB	4-6-2	1946
Ellerman Lines	35029	SR	Eastleigh	MN	4-6-2	1949
Respite	—	NCB	Hunslet	—	0-6-0ST	1950
Mine loco (3ft gauge)	—	NCB	H/Clarke (3ft gauge)	—	0-6-0 DM	1950
—	08064	BR	Darlington	08	0-6-0 DE	1953
—	3	CEGB	RSH (7746)	—	0-6-0 DM	1954
—	03090	BR	Swindon	03	0-6-0 DM	1960
—	D2860	BR	YEC	02	0-4-0 DH	1960

NRM Store

Powered Units

1931	GPO Post Office Railway Car No 809
1972	BR APT-E prototype experimental Advanced Passenger Train
1977	Leyland Motors LEV-1 experimental railbus

Departmental Stock

c1890	GNR Loco tender No 1002
1891	NER Snow plough No DE 900566
1899	GWR Hand crane No 537
1904	MR Officers' saloon No 2234
1931	LNER Petrol-driven platelayers' trolley No 960209
1932	LMS Ballast plough brake van No 197266
1938	LMS Mobile test unit No 1, No 45053
1949	BR(LMS) Dynamometer car No 3, No 45049
1957	BR Track recording trolley No DX 50002
1974	BR Matisa tamping machine No 74/007
Uncertain	LNWR Tender/water carrier

Passenger Stock

1834	B&WR 1st & 2nd composite
1834	B&WR 2nd class
1834	B&WR 3rd class
1851	ECR 1st class No 1
1895	Cambrian luggage composite No 238
1896	Glasgow Subway Car No 7
1897	Lynton & Barnstaple Rly brake composite No 6992
1925	LMS 3rd class vestibule No 7828
1925	LMS 3rd class vestibule No 7863
1927	LMS 3rd class vestibule No 8207
1936	LNER 3rd class open No 13251
1936	LNER 3rd class open No 13254
1950	BR(LMS) Corridor 3rd class brake No 27093
1955	BR Lavatory composite No E43046
1956	BR 2nd class open No E4286
1960	BR Griddle car No Sc1100
1960	Pullman Car Co 1st class Kitchen car No 311 *Eagle*
1960	Pullman Car Co 1st class Parlour car No 326 *Emerald*
1962	BR Prototype Mk II 1st class corridor No 13252

Freight & non-passenger carrying stock

1797	Peak Forest Canal quarry truck
1816	Grantham Canal Tramway truck
1828	Two replica chaldron wagons
1838	GJR replica TPO
1850	South Hetton Colliery chaldron wagon No 1155
1889	Shell-Mex oil tank wagon No 512
1894	LSWR Brake van No 99
1895	LSWR Open carriage truck No 5830
1902	NER 20-ton wooden hopper wagon
c1900	CR Bogie trolley wagon No 300041
?	GCR Single bolster wagon No 110
?	GCR Double bolster wagon No 111
1912	LB&SCR Open wagon No 27884
1912	LSWR Gunpowder van No KDS61209
1912	NER Sand wagon No DE14974
1912	GNR 8-ton van No E432764
1914	GWR Shunters' truck No W94988
1917	Shell tank wagon No 3171 (precise identity not known)
1917	MR 8-ton open wagon (precise identity not known)
1917	GCR Van (precise identity not known)
1917	NER Van (precise identity not known)
1924	LMS Van
1924	LNER Fitted tube wagon No 181358

1926	GWR Fitted open wagon No 108246
1928	ICI Nitric acid tank wagon No 14
c1933	LMS Gunpowder van No 288824
1935	SR Bogie goods brake van No 56297
1935	GWR Motor car van No 126438
1936	LNER 20-ton goods brake van No 187774
1936	LMS Tube wagon No 499254
1937	GWR Siphon bogie milk van No 2775
1938	LMS Single bolster wagon No 722702
1940	WD Warflat No 161042
1944	LMS Lowmac No M700728
1944	GWR 13-ton open wagon No DW143698
1945	GWR Loriat P No 42367
1946	SNCF 16-ton mineral wagon No ADB192437
1946	LMS Bogie trolley wagon No 300041
1946	LNER 20-ton hopper wagon No E270919
1948	BR(SR) 12-ton shock absorbing wagon No 14036
1949	BR 40-ton Gain-A No DB996724
1949	BR Bogie bolster D No B941000
1950	BR 22-ton iron ore hopper wagon No B436275
1951	ICI Liquid chlorine tank wagon No 47484
1951	BR(SR) Show cattle wagon No S3733S
1951	BR 8-ton cattle wagon No B893343
1952	BR 30-ton bogie bolster wagon No B943139
1954	National Benzole oil tank wagon No 2022
1955	BR 16-ton mineral wagon No B234830
1955	BR 16-ton mineral wagon No B227009
1957	BR Horse box No S96369
1959	BR Fish van No B87905
1959	BR Conflat No B737725
1960	BR 25-ton Weltrol WP No ADB 900916
1961	BR Presflo cement wagon No B873368
1962	BR Speedfreight container No BA 4324B
1970	Phillips Petroleum 100-ton GLW tank wagon No PP85209
?	Yorkshire Water Authority two tip wagons (2ft gauge)

TABLE 3

Exhibits that were on display at Swindon during 1990 and not expected to be on display during 1991

Locomotives

Name	*No*	*Origin*	*Builder*	*Class*	*Type*	*Built*
Puffing Billy	—	Wylam Colliery	—	—	0-4-0	1813
—	1868	LNWR	Crewe	—	2-2-2	1845
—	1439	LNWR	Crewe	—	0-4-0ST	1865
Aerolite	66	NER	Gateshead	X1 (LNER)	2-2-4T	1969
—	1275	NER	Dubs & Co	—	0-6-0	1874
Gladstone	214	LB&SCR	Brighton	B1	0-4-2	1882
Livingstone Thompson	—	FR	Boston Lodge	—	0-4-4-0T	1885
—	1008	LYR	Horwich	—	2-4-2T	1889
—	563	LSWR	Nine Elms	T3	4-4-0	1893
—	1	NER	BTH	—	Bo-Bo electric	1902
—	251	GNR	Doncaster	C1	4-4-2	1902
City of Truro	3440	GWR	Swindon	'City'	4-4-0	1903
—	K1	TGR	B/Peacock	—	0-4-0+0-4-0	1909
—	1	NSR	Stoke	—	0-4-0	1917
—	2700	LMS	Horwich	5P4F	2-6-0	1926
King George V	6000	GWR	Swindon	'King'	4-6-0	1927
—	26020	BR	Gorton/ Metrovick	76	Bo-Bo electric	1951
—	—	Imperial Paper Mills	Barclay	—	0-4-0F	1956
—	—	Tees-Side Bridge & Engineering Co	Sentinel	—	4wTG	1957
Evening Star	92220	BR	Swindon	9F	2-10-0	1960
Western Fusilier	D1023	—	Swindon	52	C-C	1963
—	41001	BR	Crewe	Prototype HST	Bo-Bo	1972

TABLE 4

Items on Loan from the NRM

Locomotives

Original type/No/Name	*Location*	*Builder*	*Built*
Hetton Colliery 0-4-0	Beamish	G. Stephenson	1822
SDR 0-4-0 *Locomotion*	Darlington	R. Stephenson & Co	1825
Shutt End Colliery 0-4-0 *Agenoria*	Science Mus	Foster Raistrick	1829
L&MR 0-2-2 *Rocket*	Science Mus	R. Stephenson & Co	1829
L&MR 0-4-0 *Sans Pareil*	Science Mus	T. Hackworth	1829
L&MR 0-2-2 *Novelty*	Museum of Science & Technology (Manchester)	Braithwaite & Ericsson	1829
GWR 2-2-2 *North Star*	Swindon GWR Mus	R. Stephenson	1837
SDR 0-6-0 No 24 *Derwent*	Darlington Nth Rd Mus	A. Kitching	1845

Wantage Tramway 0-4-0WT No 5 *Shannon*	Didcot Rly Ctr	G. England	1857
LNWR 0-4-0ST *Pet*	Tywyn NG Museum	Crewe	1865
MR 2-4-0 No 158A	Midland Rly Ctr	Derby	1866
South Devon Rly 0-4-0WT *Tiny*	Dart Valley Rly	Sara	1868
LSWR 2-4-0WT No 0298	Dart Valley Rly	B/Peacock	1874
Hebburn Works 0-4-0ST No 2 *Bauxite*	Science Mus	B/Hawthorn	1874
NER 2-4-0 No 910	Darlington Nth Rd Mus	Gateshead	1875
NER 2-4-0 No 1463	Darlington Nth Rd Mus	Gateshead	1885
C&SL Bo electric No 1	Science Mus	B/Peacock	1890
S&MR 0-4-2WT *Gazelle*	Mus of Army Transport, Beverley	Dodman	1893
GER 2-4-0 No 490	Bressingham	Stratford	1894
GWR 0-6-0 No 2516	Swindon GWR Mus	Swindon	1897
TVR 0-6-2T No 28	Caerphilly Rly Soc	TVR	1897
LSWR 4-4-0 No 120	Swanage Rly	Nine Elms	1899
GWR 4-6-0 No 4003 *Lode Star*	Swindon GWR Mus	Swindon	1907
LT&SR 4-4-2T No 80 *Thundersley*	Bressingham	R. Stephenson	1909
Nord (France) 4-6-0 No 3.628	Nene Valley Rly	Henschel	1911
NER 0-8-0 No 901	North York Moors Rly	Darlington	1919
GCR 4-4-0 No 506 *Butler Henderson*	Great Central Rly	Gorton	1920
LNWR 0-8-0 No 485	Midland Rly Ctr	Crewe	1921
GWR 4-6-0 No 4073 *Caerphilly Castle*	Science Mus	Swindon	1923
LMS 0-6-0 No 4027	Midland Rly Ctr	Derby	1924
SR 4-6-0 No 777 *Sir Lamiel*	Humberside Loco Preservation Group, Hull	N/British	1925
SR 4-6-0 No 850 *Lord Nelson*	Steamtown	Eastleigh	1926
LMS 2-6-4T No 2500	Bressingham	Derby	1934
SR 0-6-0 No C1	Bluebell Rly	Eastleigh	1942
GWR 0-6-0PT No 9400	Swindon GWR Mus	Swindon	1947
BR 4-6-2 No 70013 *Oliver Cromwell*	Bressingham	Crewe	1951
English Electric Co-Co DE *Deltic*	Science Mus	E/Electric	1955
BR A1A-A1A No 5500	Steamtown	Brush Traction	1957

Powered Units

NER electric parcels van No 3267, Tyne & Wear Mus
LPTB driving motor car No 3327, Science Mus
GWR diesel railcar No 4, Swindon GWR Mus
LMS electric motor brake 2nd No 28361, Southport Transport Mus
LMS electric driving trailer composite No 29896, Southport Transport Mus
BR APT-P pre-production Advanced Passenger Train, Crewe Heritage Ctr

Departmental Stock

LNWR Match truck No 284235, Steamtown
LNWR Steam breakdown crane No 2987, Steamtown

Passenger Stock

SDR 1st & 2nd composite No 31, Darlington Nth Rd Mus
SDR 3rd No 179, Timothy Hackworth Mus, Shildon
ECJS Royal saloon No 396, Bressingham
1910 GCR Open 3rd class No 666, Great Central Railway
1925 GWR 3rd class dining car No 9653, Severn Valley Rly
1925 GWR 3rd class dining car No 9654, Severn Valley Rly
1928 Nord (France) 2nd class corridor No 7122, Nene Valley Rly
1930 GWR composite dining car No 9605, Science Mus (Wroughton)
1930 L&MR replica 1st class *Experience*, Liverpool Mus
1930 L&MR replica 2nd class, Liverpool Mus
1937 SR buffet car No S125295, Nene Valley Rly
1937 LMS corridor 3rd class brake No 5987, Steamtown
1941 LMS Royal saloon No 798, Glasgow Museum of Transport
1947 SR 3rd class open No 1456, Bluebell Rly

Freight & non-passenger carrying stock

1907 NER bogie van No 041273, Humberside Loco Preservation Group
1917 GWR Hydra-D No 42193, Didcot Railway Ctr
1922 LB&SCR cattle truck No 7116, Isle of Wight Steam Rly
1939 SR postal sorting van No 4920, Nene Valley Rly
1947 GWR Motorcar van No 65814, Bristol Industrial Mus
1949 BR(LMS) Postal sorting van No M30272M Birmingham Rly Mus

North Yorkshire Moors Railway

This 18-mile line runs through the picturesque North York Moors National Park and is host to an extensive collection of main line locomotives
Headquarters: Pickering station, Pickering, North Yorkshire
Telephone: Pickering (0751) 72508 or 73535
Main station: Pickering
Other public stations: Grosmont, Goathland, Newtondale Halt, Levisham
OS reference: Pickering NZ 797842, Levisham NZ 818909, Goathland NZ 836013, Grosmont NZ 828053
Car parks: Grosmont, Goathland, Levisham, Pickering
Access by public transport: BR service to Grosmont from Whitby and Middlesbrough. Bus services Malton-Pickering, York or Scarborough-Pickering and Whitby-Goathland and Pickering
Refreshment facilities: Available on most trains and at Grosmont, Goathland and Pickering
Souvenir shops: Pickering, Goathland, Grosmont
Museum: Situated at locomotive depot, Grosmont
Depot: Grosmont
Length of line: 18 miles
Passenger trains: Steam-hauled services Grosmont and Pickering. The 'North Yorkshireman' runs regularly. A GWR saloon is also available for special occasions (eg wedding parties, conferences etc)
Period of public operation: Sundays (Grosmont-Goathland) January, February, March. Daily from Easter to October. Santa specials, weekends before Christmas
Facilities for disabled: The NYMR welcomes visitors who may be suffering from a disability and special attention will gladly be provided if advanced notice is given
Special notes: Operates through North York Moors National Park.
Feature article reference: RW February 1990, p93 (photo feature); RW December 1990, p749 (No 30926)

Right:
Repatriated from America and restored in Yorkshire, Southern Railway 'Schools' class 4-4-0 No 30926 *Repton* is seen undergoing final adjustments prior to its first test run on 23 August 1990.
Robin Stewart-Smith

Locomotives

Name	*No*	*Origin*	*Class*	*Type*	*Built*
Eric Treacy	45428	LMS	5MT	4-6-0	1937
*George Stephenson**	44767	LMS	5MT	4-6-0	1947
—	2238	NER	T2	0-8-0	1918
—	2392	NER	P3	0-6-0	1923
Bittern†	19	LNER	A4	4-6-2	1937
Blue Peter†	60532	LNER	A2/3	4-6-2	1948
—	2005	LNER	K1	2-6-0	1949
—	63460	NER	T3	0-8-0	1919
—	69023	LNER	J72	0-6-0T	1951
—	3814	GWR	2884	2-8-0	1940
—	6619	GWR	5600	0-6-2T	1928
—	841	SR	S15	4-6-0	1936
Repton	30926	SR	V	4-4-0	1934
—	75014	BR	4MT	4-6-0	1951
—	80135	BR	4MT	2-6-4T	1956
—†	92134	BR	9F	2-10-0	1957
Dame Vera Lynn	3672	MoS	WD	2-10-0	1943
Greyhound	D821	BR	42	B-B	1960
*Western Champion**	D1015	BR	52	C-C	1962
—	D2207	BR	04	0-6-0DM	1953
—	D5032	BR	24	Bo-Bo	1959
—	D7029	BR	35	B-B	1962
—	D8568	BR	17	Bo-Bo	1963
—	25191	BR	25	Bo-Bo	1965
—	25278	BR	25	Bo-Bo	1965
Alycidon†	55009	BR	55	Co-Co	1961

*Not on site, stored at Swindon
†Not on site

Industrial locomotives

Name	*No*	*Builder*	*Type*	*Built*
—	29	Kitson (4263)	0-6-2T	1904
—	5	R/Stephenson (3377)	0-6-2T	1909
Stanton No 44	—	Yorkshire (2622)	0-4-0DE	1956

No 21	—	Fowler (4210094)	0-4-0DH	1954
Antwerp	—	Hunslet (3180)	0-6-0ST	1944
—	12139	E/Electric (1553)	0-6-0DE	1948
—	16	Drewry	0-4-0DM	1941
—	2	R/Hornsby (421419)	4wDM	—
—	3*	R/Hornsby (441934)	4wDM	1960
—	1	Vanguard (129V)	0-4-0DM	1963
—	2	Vanguard (131V)	0-4-0DM	1963

*On loan to Middleton Railway

Stock
There are 56 coaching stock vehicles on the line comprising 32 Mk 1s, 3 Pullmans and the remainder of 'vintage' category. The 7 pre-Grouping vehicles include NER, H&BR, GNR and GCR examples. There are 14 LNER vehicles including 7 of Edward Thompson design — the largest collection in the country of this CM&EE. The 38 freight vehicles are on the NYMR include several rare or unique examples, including complete breakdown train.

Owners
841 (Essex Locomotive Society)
2005, 2238, 2392, 69023 (the North Eastern Locomotive Preservation Group)
19 and 60532 (on loan to the North Eastern Locomotive Preservation Group)
63460 (on loan from the National Railway Museum)
75014 (75014 Locomotive Operators Group)
92134 (the Standard Nine Loco Group)
Antwerp (the National Coal Board)
D821, D1015, D7029, D8568 (Diesel Traction Group)
55009 (the Deltic Preservation Society)
21, 44 (British Steel Corporation)
29 (Lambton 29 Syndicate)
926, 3814 and 80135 (private)

South Yorkshire Railway SC

Location: Barrow Road Railway Sidings, Barrow Road, Meadowhall, Wincobank, Sheffield S9 1LA
OS Reference: SK389922
Operating Society: South Yorkshire Railway
Telephone: 0742 424405 or 0742 451214
Car Park: Car parking facilities available, more facilities being developed nearby
Access by public transport: From Sheffield: By bus, No 93 Firth Park, alight at the top of Barrow Road, and follow the signposts.
From Rotherham, Doncaster and Sheffield No X77 bus alighting on Barrow Road.
By rail: Meadowhall, 100yd
Refreshment facilities: Light refreshments available
Other on site facilities: Small shop
Length of line: 3½ miles in total, although only ¾-mile is currently occupied
Public opening: Every weekend and Bank Holiday throughout the year
Period of public operation: Not operating a service as yet
Journey time: See above

Locomotives

Name	*No*	*Origin*	*Class*	*Type*	*Built*
—	D2953	BR	01	0-4-0DM	1956
—	D2854	BR	02	0-4-0DH	1960
—	D2070	BR	03	0-6-0DM	1959
—	D2199	BR	03	0-6-0DM	1961
—	D2229	BR	04	0-6-0DM	1955
—	D2284	BR	04	0-6-0DM	1960
—	D2334	BR	04	0-6-0DM	1961
Dorothy	D2337	BR	04	0-6-0DM	1961
—	D2420	BR	06	0-4-0DM	1958
—	07001	BR	07	0-6-0DE	1962
Gwyneth	D3019	BR	08	0-6-0DE	1953
—	D3476	BR	10	0-6-0DE	1957
Christine	D4092	BR	10	0-6-0DE	1962
—	12074	BR	11	0-6-0DE	1950
—	12088	BR	11	0-6-0DE	1951
Andania	40013	BR	40	1Co-Co1	1959

Industrial locomotives

Name	*No*	*Builder*	*Type*	*Built*
—	2	Hunslet (3183)	0-6-0ST	1944
—	8	Hunslet (3192)	0-6-0ST	1944
—	7	H/Clarke (1689)	0-4-0ST	1937
Cathyrn	—	H/Clarke (1884)	0-4-0ST	1955
George	—	Sentinel (9596)	4wVBT	1955
Ken	67	Sentinel (10180)	0-6-0DH	1964
Bigga	—	Fowler (4200019)	0-4-0DH	1947
Rotherham	2	YEC (2480)	0-4-0DE	1950

Membership details: 11 Linnet Mount, Thorpe Hesley, Rotherham S61 2TR

—	—	Hibbard (3817)	0-4-0DM	1956
—	44	Hunslet (6684)	0-6-0DH	1968
—	48	Hunslet (7279)	0-6-0DH	1972
—	47	T/Hill (249V)	0-6-0DH	1974
Speedy	—	Barclay (361)	0-4-0DM	1942
Hotwheels	—	Barclay (422)	0-6-0DM	1958
Toffo	2	R/Hornsby (432479)	4wDM	1959
BP21	—	H/Clarke (D707)	0-6-0DM	1950
Enterprise	—	H/Clarke (D810)	0-6-0DM	1953
—	406	NBL (277427)	0-4-0DH	1955

Rolling stock
2 diesel-electric cranes
4 Mk 1 coaches
6 Mk 1 General Utility Vehicles
1 Mk 1 bogie van
1 Covered Carriage Truck
4 LNER brake vans
Several other wagons

Stephenson Rly Museum & North Tyneside Steam Rly M

A display in buildings which began life as the Tyne-Wear Metro Test Centre now features locomotives and exhibitions which illustrate railway development from waggonways to the present day
Location: Middle Engine Lane, North Tyneside
OS reference: NZ 396576
Operating society/organisation: Tyne & Wear Museums operates Stephenson Railway Museum and North Tyneside Steam Railway Association operates North Tyneside Steam Railway for North Tyneside Council, address c/o Stephenson Railway Museum, Middle Engine Lane, West Chirton, North Shields NE29 8DX
Car park: On site
Length of line: North Tyneside Steam Railway, 1¾ miles, Stephenson Railway Museum to Percy Main
Access by public transport: Tyne & Wear Metro to Percy Main when North Tyneside Steam Railway is in operation. Bus routes 300/307 (operated by Go-Ahead Northern Co) from Newcastle (Eldon Square or Worswick Street) to North Shields, Tynemouth, Whitley Bay, stops at Museum

Locomotives

Name	*No*	*Origin*	*Class*	*Type*	*Built*
—	03078	BR	03	0-6-0DM	1959
—	12098	BR	11	0-6-0DE	1951
*Bittern**	4464	LNER	A4	4-6-2	1937

*Externally restored as 2509 *Silver Link* of 1935

Industrial locomotives

Name	*No*	*Builder*	*Type*	*Built*
Billy	—	Killingworth or RS & Co (1)	0-4-0	1816-26
—	A5	Kitson (2509)	0-6-0PT	1883
—	E4	Siemens	Bo-BoWE	1909
—	401	Bagnall (2994)	0-6-0ST	1950
—	10	Consett (10)	0-6-0DM	1958

Stock
1 NER electric parcels van
1 LNER Gresley BFK
3 BR non-gangwayed coaches

Owners
12098 (private)
Bittern (on loan from NELPG)
NER van (National Railway Museum)

Public opening: North Tyneside Steam Railway to open 5 May 1991 operating Sundays 13.00 to 17.00 to 29 September 1991. Museum displays open Wednesday to Sunday and Bank Holidays 8 May-27 October 1991
Special notes: Stephenson Railway Museum and North Tyneside Steam Railway share facilities in buildings Managed by Tyne & Wear Museums at Middle Engine Lane. North Tyneside Steam Railway Association operates and maintains exhibits from the Museum Collection
Facilities for disabled: Access for wheelchairs to Museum building at Middle Engine Lane

Tanfield Railway

The oldest railway in the world, having opened in 1725, the site maintains its basic appearance of an industrial railway but contains an unusual assortment of stock as well as the Tyneside Locomotive Collection

Industrial locomotives

Name	*No*	*Builder*	*Type*	*Built*
—	9	AEG (1565)	4w-4wE	1913
Gamma	—	Bagnall (2779)	0-6-0ST	1945
Horden	—	Barclay (1015)	0-6-0ST	1904

Location: Off the A6076 Sunniside to Stanley road
OS reference: NZ 207573
Operating society/organisation: The Tanfield Railway, Marley Hill Engine Shed, Sunniside, Gateshead, Tyne & Wear. Postal address: 33 Stocksfield Avenue, Newcastle-upon-Tyne NE5 2DX
Telephone: Newcastle (091) 2742002 (evenings)
Car park: Marley Hill Engine Shed
Access by public transport: Nearest bus stop is Sunniside, 1-mile from shed; buses 644/705/706/707/708/771/772/X11/X30/X36 from Newcastle. Summer Sundays X75 direct to Marley Hill, alight at Andrewshouse
Catering facilities: Basic refreshment facilities available on operating days
On site facilities: Shop and toilets
Length of line: 1-mile with construction of a 1-mile extension in progress which it is hoped to open during 1991
Public opening: Marley Hill Engine Shed now open daily throughout the year for viewing, except Christmas and New Years day. Steam trains run bank holiday Sundays and Mondays Easter through to August. Also every Sunday in June, July, August, September and October. Restricted diesel service other Sundays. Santa steam trains, Saturdays and Sundays in December, up to Christmas and Boxing Day. Additional services will run in 1991 on Thursdays and Saturdays. Please ring for details
Special events: Please see press for details

—	6	Barclay (1193)	0-4-2ST	1910
—	17	Barclay (1338)	0-6-0T	1913
—	32	Barclay (1659)	0-4-0ST	1920
—	3	E. Borrows (37)	0-4-0WT	1898
Wellington	—	B/Hawthorn (266)	0-4-0ST	1873
Enterprise	—	R&W Hawthorn (2009)	0-4-0ST	1884
Cyclops	—	H/Leslie (2711)	0-4-0ST	1907
—	2	H/Leslie (2859)	0-4-0ST	1911
Stagshaw	—	H/Leslie (3513)	0-6-0ST	1923
—	3	H/Leslie (3575)	0-6-0ST	1923
—	3	H/Leslie (3746)	0-6-0F	1929
Irwell	—	H/Clarke (1672)	0-4-0ST	1937
—	38	H/Clarke (1823)	0-6-0T	1949
—	4	Sentinel (9559)	0-4-0T	1953
—	169	RSH (6980)	0-4-0DM	1940
Hendon	—	RSH (7007)	0-4-0CT	1940
—	62	RSH (7035)	0-6-0ST	1940
—	3	RSH (7078)	4w-4wE	1940
—	49	RSH (7098)	0-6-0ST	1943
Progress	—	RSH (7298)	0-6-0ST	1946
Cochrane	—	RSH (7409)	0-4-0ST	1948
—	44	RSH (7760)	0-6-0ST	1953
—	38	RSH (7763)	0-6-0ST	1954
—	21	RSH (7796)	0-4-0ST	1954
—	47	RSH (7800)	0-6-0ST	1954
—	1	RSH (7901)	0-4-0DM	1958
—	16	RSH (7944)	0-6-0ST	1957
FGF	—	Barclay (D592)	0-4-0DH	1969
—	2	A/Whitworth (D22)	0-4-0DE	1933
Escucha	11	B/Hawthorn (748)	0-4-0ST	1883*

*600mm gauge

Stock

16 4-wheel carriages
3 6-wheel carriages
1 6-wheel van
14 hopper wagons
9 contractors bogies
3 brake vans
3 steam cranes
7 covered wagons
4 open wagons
4 black wagons
4 flat wagons

Facilities for disabled: Good access to stations and coaches
Special notes: The railway is extending its line back to East Tanfield, and also features the Tyneside Locomotive Collection.
Feature article reference: RW January 1990, p29 (photo feature)
Membership details: Miss E. Martin, 33 Stocksfield Avenue, Fenham, Newcastle-upon-Tyne NE5 2DX

Vintage Carriage Museum — M

A unique collection of elderly railway carriages. These include a fully restored and prize-winning 1876-built Manchester, Sheffield & Lincolnshire Railway tricomposite coach (in Great Central Railway livery), together with fully-restored Metropolitan Railway and South Eastern & Chatham Railway coaches and other coaches in various stages of restoration.

Audio presentations ('Travellers' Tales') and access to coach interiors bring the collection to life and allow close examination of the restoration work, completed and in progress. Other railway items on display include a fine collection of Frank Newbould and other railway posters. A veteran

Stock

Railway	*BR or previous owner number*	*Type*	*Date built*	*Seats*	*Weight*	*Length*
MS&LR	176	4-wheel 1st/2nd/3rd/ luggage	1876	34	12T	28ft 0in
GNR	?	6-wheel 3rd brake	1888	40	14T	34ft 11in
MR	358	6-wheel 1st/3rd/ luggage	1886	32	15T	34ft 0in
Met	427	BS	1910	84	30T	54ft 0in
Met	465	S	1919	108	30T	54ft 0in
Met	509	F	1923	84	30T	54ft 0in
SR (SECR)	S3554S	BSK	1924	42	33T	65ft 3in
BR (SR)	S1469S	TSO	1951	64	32T	67ft 1in

(Also two oil tank wagons: MR, c1890, and Esso, 1939)

small steam locomotive is usually also present, with access to the footplate
Location: Vintage Carriage Museum, Ingrow Station Yard, South Street, Keighley, West Yorkshire
Operating society/organisation: Vintage Carriages Trust (a Registered Charity), c/o The Railway Station, Haworth, Keighley, West Yorkshire BD22 8NJ
Telephone: Keighley (0535) 680425
Car Park: Yes. Also limited coach parking at Ingrow station
Access by public transport: By Rail to Keighley (one mile), or by Worth Valley Railway train to Ingrow West (adjacent): good bus service from Keighley town centre
On site facilities: Toilets, small sales stall
Public opening: 12.00-16.00 whenever Worth Valley Railway trains are running. Special openings outside these times and dates, and school and other party bookings, are welcome: telephone 0535 680425 or 0535 46472
Facilities for the disabled: Limited wheelchair access
Special notes: Most, but not all, of the collection is displayed in the Museum at any one time. One or more of the carriages (and also the locomotive *Bellerophon*) are occasionally in passenger use on the Worth Valley Railway or elsewhere

Industrial locomotives

Name	*No*	*Builder*	*Type*	*Built*
Bellerophon	—	Haydock Foundry (C)	0-6-0WT	1874
Sir Berkeley	—	M/Wardle (1210)	0-6-0ST	1891
Lord Mayor	—	H/Clarke (402)	0-4-0ST	1893

Membership details: Membership Secretary c/o above address (Haworth station)

West Yorkshire Transport Museum M

The West Yorkshire Transport Museum's collection consists of some 70 items, of which 13 are rail or tramway vehicles. It is hoped to progressively reopen the Spen Valley Railway, commencing at Low Moor. The railway would be all electric, traction being by integral battery, or overhead electric supply.

Two locomotives and five trams are to be seen at the Ludlam Street Depot. Also stored at Ludlam Street are about 50 road vehicles, mostly motorbuses, but including five tolleybuses and service vehicles.

Recently approved plans will see the opening of the West Yorkshire Transport Centre at Low Moor, Bradford, in the spring of 1992. Proposed attractions include the operation of trams, trolleybuses and rail vehicles backed by extensive interpretation of the West Yorkshire Transport story
Location: Ludlam Street Depot, Mill Lane, off Manchester Road, Bradford, 400yd south of Bradford Interchange, just off A641
OS reference: SE 164322
Operating society/organisation: West Yorkshire Transport Museum Trust Ltd
Telephone: Bradford (0274) 736006
Car park: Yes, free
Access by public transport: Rail/road: 10min walk from Bradford Interchange station, 2min walk from nearest Manchester Road bus stops

Locomotives

Name	*No*	*Origin*	*Class*	*Type*	*Built*
Electra	27000	BR	EM2	Co-Co	1953

Industrial locomotives (all electric)

Name	*No*	*Builder*	*Type*	*Built*
—	10	Siemens	Bo	1913
—	1	E/Electric	Bo	1935

Stock
Derby 'Lightweight' Battery Electric unit Sc79998/9 (on loan to East Lancashire Railway)
Class 506 Electric Unit — 59404/504/604 (on loan to Midland Railway Centre)
1 Blackpool tram 663
1 Rotterdam tram 408
2 Budapest trams 2576/2577
1 Brussels tram 96
Various tram bodies

On site facilities: Viewing of motorbuses, trolleybuses and trams at Ludlam Street; rides on preserved motorbuses are available; souvenir shop and light refreshments available. Special events as advertised
Public opening: Ludlam Street is open every Sunday, from 11.00-17.00. Parties welcome at other times by prior arrangement. Details of special events, etc, may be obtained by telephoning the museum.
Facilities for disabled: No special toilet facilities, but all parts of museum accessible by wheelchairs
Membership details: The West Yorkshire Transport Museum Society provides volunteers who support the project in many ways, including the restoration of Trust-owned exhibits. Telephone the museum for details

Scotland

Bo'ness & Kinneil Railway SC

A few miles west of Edinburgh, this railway has won several awards, particularly for its recreated station buildings — the entire railway has been constructed on a 'green field site'. Achieved in 1989 was the extension of the operating line inland to Birkhill where the railway provides access to a new attraction, the Birkhill Clay Mine

Location: Scottish Railway Preservation Society, Bo'ness Station, Union Street, Bo'ness, West Lothian, EH51 0AD

Access by public transport: Nearest BR stations — Falkirk Grahamston, Linlithgow, Dalmeny

OS reference: NT 003817

Telephone: Bo'ness (0506) 822298

Main station: Bo'ness

Car park: At the site (free)

Refreshment facilities: Bo'ness 'Bufferstop' cafe with snacks, unlicensed

Souvenir shop: Bo'ness and secondhand sale first Sunday throughout the year

Depot: Bo'ness

Length of line: The line was extended to 3½ miles (to Birkhill) from Easter 1989

Passenger trains: Trains run at 11.05 (July and August), 12.10 (Saturdays excepted April and May), 13.15, 14.20, 15.25 and 16.30 in season on Saturdays and Sundays. Note: 16.30 diesel-hauled except weekdays July and August

Period of public operation: 30 March-13 October. Bank holiday Mondays 1 April, 6, 27 May, 16 July. Daily 15 July-18 August. Schools services late May/early June. Pre-Christmas.

Special events: Santa Specials 7/8, 14/15, 21/22 December

Facilities for disabled: At Bo'ness for shop and cafeteria

Feature article reference: RW September 1989, p528

Locomotives

Name	*No*	*Origin*	*Class*	*Type*	*Built*
Sovereign	44871	LMS	5MT	4-6-0	1945
—	419	CR	439	0-4-4T	1908
—	1313	SJ	B	4-6-0	1917
Morayshire	246	LNER	D49	4-4-0	1928
Maude	673	NBR	—	0-6-0	1891
—	80105	BR	4MT	2-6-4T	1955
—	8 (D9524)	BR	14	0-6-0DH	1964
—	25235 (D7585)	BR	25	Bo-Bo	1965
—	D5351	BR	27	Bo-Bo	1961

Industrial locomotives

Name	*No*	*Builder*	*Type*	*Built*
Clydesmill	3	Barclay (1937)	0-4-0ST	1928
—	3	Barclay (2046)	0-4-0ST	1937
—	24	Barclay (2335)	0-6-0T	1953
Texaco	—	Fowler (4210140)	0-4-0DM	1958
Lord King	—	H/Leslie (3640)	0-4-0ST	1926
—	19	Hunslet (3818)	0-6-0ST	1954
DS3	—	R/Hornsby (275883)	4wDM	1949
DS4	P6687	R/Hornsby (312984)	0-4-0DE	1951
(Ranald)	—	Sentinel (9627)	4wVBT	1957
—	970214	Wickham (6050)	2w-2PMR	c1951
—	—	Matisa (48626)	—	—
—	5	Hunslet (3837)	0-6-0ST	1955
—	(7)	Bagnall (2777)	0-6-0ST	1945
Borrowstounness	—*	Barclay (840)	0-4-0T	1899
—	—*	M/Rail (110U082)	4wDH	1970
—	—	Wickham (10482)	2w-2PMR	1970
—	(17)	Hunslet (2880)	0-6-0ST	1943
—	970213	Wickham (6049)	2w-2PMR	c1951
—	17	Barclay (2296)	0-4-0ST	1952
Lady Victoria	3	Barclay (1458)	0-6-0ST	1916
The Wemyss Coal Co Ltd	20	Barclay (2068)	0-6-0T	1939
—	(6)	Barclay (2127)	0-4-0CT	1942
No 1	—	Barclay (343)	0-6-0DM	1941
City of Aberdeen	—	B/Hawthorn (912)	0-4-0ST	1887
F82 (Fairfield)	—	E/Electric (1131)	4wBE	1940
(Ellesmere)	—	Hawthorns (Leith) (244)	0-4-0WT	1861
Kelton Fell	13	Neilson (2203)	0-4-0ST	1876
—	(1)	Neilson Reid (5710)	0-6-0T	1902
Kilbagie	DS2	R/Hornsby (262998)	4wDM	1949
DS6	(1)	R/Hornsby (421439)	0-4-0DE	1958
St Mirren	(3)	R/Hornsby (423658)	0-4-0DE	1958

—	D88/003	R/Hornsby (506500)	4wDM	1965
—	(DS5)	R/Hornsby (423662)	0-4-0DE	1958
—	—	BTH	4wE	1908
(Denis)	—	Sentinel (9631)	4wVBT	1958
—	—	Arrols (Glasgow)	2w-2DM	c1966

*3ft 0in gauge

Stock
A large selection of coaching stock, many built by Scottish pre-Grouping companies, and an appropriate collection of early freight vehicles

Owners
80105 and (*Denis*) (owned by Scottish Locomotive Owners Group)
246 and 24 (owned by Royal Museum of Scotland)

Caledonian Railway (Brechin) SC

Information enquiries: Iain A. H. Smith, Publicity Director, C/o 30 Moathill East, Cupar, Fife KY15 4DT
Headquarters: Caledonian Railway (Brechin) Ltd, The Station, 2 Park Road, Brechin, Angus DD9 7AF
Telephone: (0334) 55965, after 4.30pm Mon-Fri or George Mitchell (03562) 4562 after 4.30pm
Main station: Brechin
Other public stations: Bridge of Dun
OS reference: NO 603603
Car park: Brechin, Bridge of Dun
Access by public transport: Nearest BR station — Montrose, 9 miles. Local bus operator, Strathtay Scottish
Refreshment facilities: Light refreshments available during steamdays at Brechin
Souvenir shop: Brechin
Museum: Brechin
Length of line: 4 miles
Depot: Brechin
Passenger trains: It is hoped that by May 1991 a steam service will be operated between Brechin and Bridge of Dun (see railway press for details)
Period of public operation: Sundays only 5 May-29 September
Special events: 31 March — Easter Specials (shuttle service, 1-mile); 21 July — Gala Day; 27 October — Haloween Specials; 8, 15, 22 December — Santa Specials; 26 December — Boxing Day Specials. Other events may be planned, see local press for details
Facilities for disabled: Easy access to both stations, access to toilets and shop, passengers can be assisted on and off trains by railway staff
Disclaimer: The Caledonian Railway (Brechin) Ltd, reserve the right to amend, cancel or add to these events. And whilst every effort will be made to maintain the above services, the company does not guarantee that trains will depart or arrive at the time stated and reserves the right to suspend or alter any train without notice and will not accept any liability for loss, inconvenience or delay thereby caused
Membership details: Murray D. Duncan, 2 Binghill Crescent, Milltimber, Aberdeen

Locomotives

Name	*No*	*Origin*	*Class*	*Type*	*Built*
—	46464	LMS	2MT	2-6-0	1950
—	D2866	BR	02	0-4-0DM	1961
Brechin City	D3059	BR	08	0-6-0DE	1954
—	27001	BR	27	Bo-Bo	1961

Industrial locomotives

Name	*No*	*Builder*	*Type*	*Built*
—	6	Bagnall (2749)	0-6-0ST	1944
Bon Accord	2	Barclay (807)	0-4-0ST	1897
—	1	Barclay (1863)	0-4-0ST	1926
Diana	1	Hunslet (2897)	0-6-0ST	1943
Yard No DY326	—	Hibberd (3743)	4wDM	1955
—	2/88	R/Hornsby (275880)	4wDM	1949
—	6	R/Hornsby (421700)	4wDM	1959

Stock
3-car Class 126 diesel multiple-unit, Nos 51017, 51043, 59404
3 ex-BR Mk 1 TSOs
1 ex-BR Mk 1 BSK
1 ex-BR Suburban
1 ex-BR Engineers Inspection Saloon
1 ex-BR Mk 1 sleeping car
1 ex-BR Full Kitchen Car
Various items of freight stock

Owners
Bon Accord (Aberdeen Town Council)
Class 126 DMU (the Swindon Preservation Society)
27001 (the Class 27 Locomotive Group)

Glasgow Museum of Transport M

Glasgow's magnificent railway collection represents one of the best efforts by a municipal authority to preserve a representative collection of items appropriate to the 'locomotive builders to the Empire'. In 1989 the collection opened to view once again in its new setting at the former Kelvin Hall

Access by public transport: Strathclyde PTE Underground Kelvinhall: Strathclyde Buses 6, 6A, 8, 8A, 9, 9A, 16, 42, 42A, 44A, 57, 57A, 62, 62A, 62B, 64; Kelvin Scottish Buses 5, 5A; Clydeside Scottish Buses 17

Operating society/organisation: Glasgow City Council, Museums & Art Galleries Department

Location: Museum of Transport, Kelvin Hall, 1 Bunhouse Road, Glasgow G3 8DP

Telephone: 041-357-3929

Car park: Opposite Museum entrance

On site facilities: Toilets, cafeteria, shop and public telephone

Locomotives

Name	*No*	*Origin*	*Class*	*Type*	*Built*
—	123	CR	123	4-2-2	1886
—	9	G&SWR	5	0-6-0T	1917
—	103	HR	—	4-6-0	1894
Glen Douglas	256	NBR	K	4-4-0	1913
Gordon Highlander	49	GNSR	F	4-4-0	1920

Industrial locomotives

Name	*No*	*Builder*	*Type*	*Built*
—	1	Barclay (1571)	0-6-0F	1917
—	—	Chaplin (2368)	0-4-0TG	1888
—	—	BEV (583)	B	1927

Stock

Glasgow District Subway car 39T
Glasgow Corporation Underground cars 1 and 4
LMS King George VI's saloon 498 of 1941

Public opening: Monday-Saturday 10.00-17.00; Sunday 14.00-17.00. Closed 1 January and 25 December only

Facilities for disabled: Access to all areas but entry via separate entrance — enquire at main door

Feature article reference: RW March 1989, p155 (locomotive 123)

Lochty Railway SC

Headquarters: 23 Millfield, Cupar, Fife KY15 5UT

OD reference: NO 522080

Telephone: 0334 54815 (Mon-Sat only)

Main station: Lochty

Car park: Lochty

Access by public transport: 10 miles from Cupar, 7 miles from Crail (B940). BR 10 miles from Cupar, then taxi; Kirkcaldy 18 miles, bus to Crail, then taxi

Souvenir shop: Lochty

Passenger trains: Steam-hauled from Lochty through the estate of J. B. Cameron (1½ miles)

Period of public operation: 16 June-8 September. Lochty station site open from 12.00. Regular passenger trains 13.45-17.00

Special notes: This line was opened as Scotland's first preserved railway in 1967. No statutory authority was required to open the line, now in its 25th year

Membership details: c/o above address

Feature article reference: RW March 1990, p140 (Locomotive 60009)

Locomotives

Name	*No*	*Origin*	*Class*	*Type*	*Built*
Osprey	60009	LNER	A4	4-6-2	1937

Industrial locomotives

Name	*No*	*Builder*	*Type*	*Built*
—	1	Peckett (1376)	0-4-0ST	1915
—	LPR No 16	Bagnall (2759)	0-6-0ST	1944
North British	LPR No 4	R/Hornsby (421415)	4wDM	1958
—	NCB 21	Barclay (2292)	0-4-0ST	1951
Forth	SGB 10	Barclay (1890)	0-4-0ST	1926
—	NCB 10	N/British (27591)	0-6-0DH	1957
—	LPR 34	R/Hornsby (506399)	0-4-0DH	1964
River Eden	RAF 400	N/British (27421)	0-4-0DH	1955
River Tay	RAF 405	N/British (27426)	0-4-0DH	1955

Locomotive notes: SGB 10 is the usual service locomotive.

Note: No 60009 *Osprey* (formerly *Union of South Africa*) is stored at Markinch station for main line use on railtours.

Stock

3 ex-BR Mk 1 coaches
1 ex-BR Shark ballast brake van
1 LNER observation coach
1 ex-NBR 8-ton box van (built 1901)
13 assorted private owner wagons
1 flat wagon

Mull Rail

Operating society/organisation: Mull & West Highland (NG) Railway Co Ltd, Old Pier Station, Craigmure, Isle of Mull PA65 6A7
Telephone: 06802 494
OS reference: Sheet 49, 725369
Car park: At Craigmure
Access by public transport: Caledonian Macbrayne ferry from Oban
On site facilities: Gift shop
Length of line: 1¼ miles/10¼in gauge
Public opening: Easter, then daily 29 April-12 October
Facilities for disabled: No steps on railway, two compartments for wheelchairs

Locomotives

Name	*No*	*Builder*	*Type*	*Built*
Lady of the Isles	—	Marsh	2-6-2T	1981
Waverley	—	Curwen	4-4-2	1948
—	—	Alcock	4w-4PM	1973
Glen Auldyn	—	Davies	8DH	1986

Rolling stock
11 coaches (two with wheelchair accommodation)
3 bogie wagons
1 4-wheel wagon

Membership details: Friends of Mull Rail, M. R. Oliver, Kilmory Drive, Lochgilphead, Argyll

Paddle Steamer Preservation Society

The PS *Waverley* was built to the order of the London & North Eastern Railway in 1946; she replaced a vessel of the same name which was sunk off Dunkirk during May 1940. Sold to the PSPS in 1974 she now offers cruises around the coast of the United Kingdom in company with MV *Balmoral*. Also in the 'fleet' is PS *Kingswear Castle* which cruises on the Thames and Medway
Headquarters: Waverley Excursions Ltd, Waverley Terminal, Anderston Quay, Glasgow G3 8HA
Telephone: 041 221 8162
On ship facilities: All vessels have catering facilities, bars, toilets (disabled only on *Waverley*)

Membership details: Paddle Steamer Preservation Society, PO Box 385, Hazlemere, High Wycombe HP11 1AE

Details of all the cruises operated by the three vessels can be obtained from the above address

Scottish Industrial Railway Centre SC

The Scottish Industrial Railway Centre is based on part of the former Dalmellington Iron Co railway system which was one of the best known industrial railway networks in Britain. Steam worked up until 1978 when the system closed and it is the aim of the centre to recreate part of the railway. The Ayrshire Railway Preservation Group also own the former G&SWR station at Waterside, 2 miles from the centre, and have access to the former NCB locomotive shed and wagon workshops at Waterside. These locations are not yet open to the general public. Working in conjunction with the Dalmellington & District Conservation Trust it is hoped to create an industrial heritage centre at

Locomotives

Name	*No*	*Origin*	*Class*	*Type*	*Built*
—	MP228 (12052)	BR	11	0-6-0DE	1949
—	MP229 (12093)	BR	11	0-6-0DE	1951

Industrial locomotives

Name	*No*	*Builder*	*Type*	*Built*
—	16	A/Barclay (1116)	0-4-0ST	1910
—	8	A/Barclay (1296)	0-6-0T	1912
—	19	A/Barclay (1614)	0-4-0ST	1918
Aberdeen Corporation Gas Works	3	A/Barclay (1889)	0-4-0ST	1926
—	8	A/Barclay (1952)	0-4-0F	1928
—	10	A/Barclay (2244)	0-4-0ST	1947
—	25	A/Barclay (2358)	0-6-0ST	1954
—	1	A/Barclay (2368)	0-4-0ST	1955
—	118	A/Barclay (366)	0-4-0DM	1940
—	7	A/Barclay (399)	0-4-0DM	1956

Waterside based on the iron, coal and brick making industries
Location: Scottish Industrial Railway Centre, Minnivey Colliery, Dalmellington, Ayrshire
OS reference: NS 476074
Operating society/organisation: Ayrshire Railway Preservation Group
Telephone: The Secretary (0292) 313579
Length of line: ¼-mile
Access by public transport: Nearest BR station, Ayr (14 miles). Western Scottish bus service from Ayr, Nos 52/53. Tel: (0292) 264643
On site facilities: Steam-hauled brake van rides. Guided tours of centre, museum of railway relics and photographs, souvenir shop and buffet
Public opening: Open for static display with limited facilities every Saturday, June to end September
Special events: Steam Open Days 1991: 26 May; 29/30* June; 14, 27/28* July; 14, 24/25* August; 29 September. *Fireless locomotive expected to be in steam. Opening times: 11.00-16.00
Membership details: Mr Frank Beattie, 1 McKnight Avenue, Waterside, Fenwick, Kilmarnock, Ayrshire

Lily of the Valley	—	Fowler (22888)	0-4-0DM	1943
—	—	R/Hornsby (224352)	—	—
Blinkin Bess	—	R/Hornsby (284839)	4wDM	1950
Johnnie Walker	—	R/Hornsby (417890)	4wDM	1959
—	—	R/Hornsby (421697)	0-4-0DM	1959
—	107	Hunslet (3132)	0-4-0DM	1944
—	—	Sentinel (10012)	4wDM	1959
—	—	Donnelli (163)	4wDMR	1979
(3ft gauge)				
—	—	R/Hornsby (256273)	4wDM	1949
—	—	Hunslet (8816)	4wDH	1981
(2ft 6in gauge)				
—	2	R/Hornsby (183749)	4wDM	1937
—	3	R/Hornsby (210959)	4wDM	1941
—	1	R/Hornsby (211681)	4wDM	1942

Note: Not all vehicles on public display

Stock
2 Wickham trolleys
1 steam crane
Various other items

Right:
The Scottish Industrial Railway Centre at Dalmellington, Ayrshire: NCB Ayrshire Area No 10 (Barclay No 2244 of 1947) hauling a short demonstration freight train on 27 May 1990.
Gordon Thomson

Scottish Mining Museum — Prestongrange SC

Location: On the B1348 between Musselburgh and Prestonpans.
OS reference: NT 734737
Operating society/organisation: Scottish Mining Museum Trust, Lady Victoria Colliery, Newtongrange, Midlothian EH22 4QN
Telephone: 031-663-7519
Car park: On site
On site facilities: Visitor Centre — 'Pots at the "Pans"', an exhibition of the potteries of Prestonpans, Musselburgh and Portobello. Machine Exhibition Hall, Cornish Beam Pumping Engine installed in 1874. In the David Spence Gallery, the former Power House, 'Cutting the Coal', an

Industrial locomotives

Name	*No*	*Builder*	*Type*	*Built*
—	6	A/Barclay (2043)	0-4-0ST	1937
—	17	A/Barclay (2219)	0-4-0ST	1946
Prestongrange	7	G/Ritchie (536)	0-4-2ST	1914
Tomatin	1	M/Rail (9925)	4wDM	1963
—	—*	Hunslet (4440)	4wDM	1952
—	32	R/Hornsby	4wDM	
—	33	R/Hornsby	4wDM	

*2ft gauge

Rolling stock
Steam crane, Whittaker No 30, c1890, occasionally in steam

Locomotive notes: *Prestongrange* is expected to be out of service during 1991, 6 to be available for passenger trains

exhibition which traces the changing methods of cutting coal. There is an underground gallery, a coalface, a reconstruction of a colliery workshop and a magnificent collection of coal cutting machines and equipment
Toilets: Visitor Centre
Refreshment facilities: Available at Visitor Centre
Public opening: Easter Sunday to the end of September, 11.00-16.00, last tour at 15.00
Length of line: 400m (standard gauge) extension in progress
Facilities for disabled: Access and toilet at Visitor Centre. Access to 'Cutting the Coal' exhibition, and footpaths along the site
Special events: Steam during summer months — date to be confirmed
Special notes: Locomotive in steam first Sunday of month April-September, passenger rides available. Advance notice is required for large parties
Membership details: Friends of the Scottish Mining Museum, Elaine Carmichael c/o above address

Strathspey Railway

Scotland's steam railway in the highlands connects the busy tourist resort at Aviemore to the more traditional highland village of Boat of Garten, famed as one of the few nesting places of the osprey (viewing site 3 miles from station). A feature of the railway is the special evening dinner and Sunday lunch trains. The former operating during June/August, the latter throughout the season
Enquiries: Aviemore Speyside Station, Dalfaber Road, Aviemore, Inverness-shire PH22 1PM
Telephone: (0479) 810725
Main station: Aviemore (Speyside)
Other public stations: Boat of Garten
OS reference: Aviemore NH 898131, Boat of Garten NH 943789
Car parks: Aviemore and Boat of Garten
Access by public transport: BR services and express bus to Aviemore. Highland buses local service to Boat of Garten
Refreshment facilities: On train buffet car or facilities on many trains. Evening and Sunday lunch time diner trains — booking essential — Tel: 047 983 258 for details/bookings
Souvenir shop: Boat of Garten and Aviemore (Speyside)
Museum: Small relics display at Boat of Garten
Depot: Aviemore (not open to public)
Length of line: 5 miles
Journey time: 17min, return within the hour possible on most services
Passenger trains: Steam- and some diesel-hauled services. Boat of Garten-Aviemore
Period of public operation: Please contact for details
Facilities for disabled: Access possible at Boat of Garten and Aviemore (Speyside). Please contact in advance for directions and if a party involved

Locomotives

Name	*No*	*Origin*	*Class*	*Type*	*Built*
—	5025	LMS	5MT	4-6-0	1934
—	46512	LMS	2MT	2-6-0	1952
—	828	CR	812	0-6-0	1899
—	08490	BR	08	0-6-0DE	1958
—	D5394	BR	27	Bo-Bo	1962

Locomotive notes: In service 5025, D5394. Under restoration 46512, 828. Under repair 08490.

Industrial locomotives

Name	*No*	*Builder*	*Type*	*Built*
—	48	Hunslet (2864)	0-6-0ST	1943
Cairngorm	9	RSH (7097	0-6-0ST	1943
—	60	Hunslet (3686)	0-6-0ST	1948
Victor	2996	Bagnall (2996)	0-6-0ST	1951
Niddrie	6	Barclay (1833)	0-6-0ST	1924
*Forth**	10	Barclay (1890)	0-4-0ST	1926
Dailuaine	1	Barclay (2073)	0-4-0ST	1939
Balmenach	2	Barclay (2020)	0-4-0ST	1936
—	17	Barclay (2017)	0-6-0T	1935
Inveresk	14	R/Hornsby (260756)	0-4-0DM	1950
Inverdon	15	Simplex (5763)	4wDM	1957
—	16	North British (27549)	0-4-0DM	1951
Queen Anne	20	R/Hornsby (265618)	4wDM	1948

Locomotive notes: In service 3, 60, 2996.

Stock
14 ex-BR coaches
5 ex-LMS coaches
1 Pullman coach
3 ex-LMS sleeping cars
1 ex-LNER sleeping car
1 ex-LNWR coach
1 ex-HR coach
1 ex-NBR coach
2 ex-GNSR coaches
Numerous examples of rolling stock
*Not on site

Owners
6, 17 and 46512 (the Highland Locomotive Co Ltd)
828 (the Scottish Locomotive Preservation Trust Fund)

Special notes: First and third class travel available on most trains. Family fares available for third class travel. Special rates/arrangements for parties.
Membership details: Strathspey Railway Association at above address

Ireland

Downpatrick Railway Project

Location: Downpatrick Station, Market Street, Downpatrick, Co Down
OS reference: J483444
Operating society/organisation: Downpatrick & Ardglass Railway Co Ltd, with the support of the Downpatrick Railway Society
Telephone: (0396) 830141 or (0396) 615779
Car park: Available adjacent to station in Downpatrick
Access by public transport: A regular service is operated by Ulsterbus from Belfast Tel: (0232) 320011
Refreshment facilities: Buffet carriage open on operating days
On site facilities: Souvenir shop, toilets
Length of line: ½-mile
Public opening: St Patricks Day (17 March); Easter Monday and Tuesday; 6, 25 May; Sundays in July; Weekends in August and Sunday 1 September
Period of public operation: 14.00-17.00
Journey time: 30min return journey to Quoile marshes
Special events: 25 May — Enthusiast's Day; 25 August — Gala Day; 27 October — Ghost Trains; Easter Egg and Santa Specials
Facilities for disabled: All station facilities at Downpatrick have facilities for disabled
Membership details: Peter King, Secretary, Downpatrick Railway Society, 15 Ravensdene Park, Belfast BT6 0DA, Northern Ireland. (0232) 643830

Locomotives

Name	*No*	*Origin*	*Class*	*Type*	*Built*
—	E421	CIE	421	C	1962
—	E432	CIE	421	C	1962
—	G613	CIE	611	B	1962

Industrial locomotives

Name	*No*	*Builder*	*Type*	*Built*
Guinness	3BG	H/Clarke (1152)	0-4-0ST	1919
—	1	O&K (12475)	0-4-0T	1934
—	3	O&K (12662)	0-4-0T	1935

Rolling stock
1 CIE Brake open standard
1 CIE Brake open standard generating steam van
1 CIE Buffet open standard
2 NCC parcels vans
1 NCC open wagon
1 GNR(I) brake van
1 CIE closed van
1 GS&WR ballast hopper
1 GSR ballast hopper
Belfast & County Down Railway 'Royal Saloon' No 153
1 B&CDR 1st/2nd composite (No 152)
1 B&CDR 3rd open (ex-railmotor)
1 B&CDR 6 wheeled brake 3rd (No 39)
1 B&CDR 6 wheeled 2nd (No 154)
1 GS&WR 3rd open (No 836)
1 GS&WR ballast plough van (on loan from Westrail [Tuam] Ltd)
1 GNR(I) 6 wheeled saloon (No unknown)

Owners
1 and 3 (the Irish Sugar Locomotive Group)
G613 (on loan from Westrail [Taum] Ltd)
3BG (on loan from the Railway Preservation Society of Ireland)

Irish Steam Preservation Society — M

Location: Stradbally Hall, eight miles from Athy, six miles from Portlaoise.
Telephone: (0502) 25136
Access by public transport: Irish Rail train to Athy or Portlaoise. Kavanagh's Bus Portlaoise-Stradbally-Athy (daily Monday-Saturday)
On site facilities: 3ft gauge railway. Museum in nearby town
Catering facilities: None on site but town centre ¼-mile away
Length of line: 1km
Public opening: Public open days 31 March/1 April — Easter; 2/3 June — June Bank Holiday; 27 July — Open Day and Flower Festival; 4/5 August — National Steam Rally; 15 September; 27/28 October. 14.30-17.00 each day (noon-18.30, 4/5 August)

Industrial locomotives

Name	*No*	*Builder*	*Type*	*Built*
—	2	Barclay (2264)	0-4-0WT	1949
—	—	Hunslet (2281)	4wDM	1941
Nippy	—	Planet (2014)	4wDM	1936
—	4	R/Hornsby (326052)	4wDM	1952

Stock
1 Passenger coach
1 Ballast wagon
1 Brake van

Railway Preservation Society of Ireland — SC

The RPSI is more famed for its activities elsewhere in Ireland than for its line based at Whitehead and Mullingar. The Society enjoys excellent co-operation from the two national railway systems of both parts of Ireland and its spring 'Three-Day Railtour' (which normally provides up to five days of activities) has become renowned world-wide and regularly sells out well in advance. Numerous other day excursions are operated throughout the year for those unable to participate in the spring weekend tour. Short steam train rides are also provided on certain dates at the main depot at Whitehead which is well worth a visit to view the Society's extensive collection. Visits by organised groups are welcome if advanced notification is given in writing to Whitehead Operations Manager at RPSI Whitehead BT38 9NA
Location: Whitehead Excursion station N. Ireland
Operating society: Railway Preservation Society of Ireland
Telephone: Whitehead (09603) 53567
Car park: At the station
Access by public transport: Northern Ireland Railways services from Belfast (York Road) or Larne Harbour: Ulster Bus service
On site facilities: Souvenir shop, refreshments in dining car, open on public operating days. Souvenir shop also open 14.00-17.00 on third Saturday of month all year, except on main line train operating days
Public opening: Easter Sunday, Monday and Tuesday, and for Santa services on December Sundays before 25 December
Special notes: The RPSI is noted for its main line railtours operated each year using primarily its own rolling stock. For details telephone above number and leave name and address or write enclosing SAE.
Special events: Easter Bunny rides at Whitehead 31 March, 1/2 April

Locomotives

Name	*No*	*Origin*	*Class*	*Type*	*Built*
Merlin	85*	GNR (I)	V	4-4-0	1932
Slieve Gullion	171	GNR (I)	S	4-4-0	1913
—	4	LMS (NCC)	WT	2-6-4T	1947
—	184†	GS&WR	J15	0-6-0	1880
—	186	GS&WR	J15	0-6-0	1879
—	461	D&SER	K2	2-6-0	1922
Lough Erne	27	SL&NCR	Z	0-6-4T	1949

Industrial locomotives

Name	*No*	*Builder*	*Type*	*Built*
Guinness	3‡	H/Clarke (1152)	0-4-0ST	1919
R. H. Smyth	3	Avonside (2021)	0-6-0ST	1928
—	23	Planet (3509)	0-4-0DM	1951
—	4	R/Hornsby	0-4-0DM	1954

*On loan from Ulster Folk & Transport Museum
†Normally based in the Republic of Ireland
‡On loan to Downpatrick & Ardglass Railway Society

Stock
The Society also owns some 20 operational coaches, normally resident at Whitehead and Mullingar. A small number of freight wagons are also preserved

(14.00-17.00); Decies Railtour 11-13 May (Dublin-Waterford-Dublin-Belfast); Atlantic Coast Express 9 June (Belfast-Londonderry-Belfast); Mid-Summer Wine & Cheese Special 22 June (Belfast-Portrush-Belfast); Portrush Flyer 20 July, 13, 17 August (Belfast-Portrush-Belfast); Special Event at Whitehead 24-26 August; Steam 'n' Jazz Evening Special 13 September (Whitehead-Belfast*); Steam Enterprise 21, 28 September (Belfast*-Dublin-Belfast*); Larne Lough 13 October (Belfast*-Whitehead); Santa Steam Rides at Whitehead 8, 15, 22 December.
Note: Belfast departures from York Road station unless otherwise indicated*, as Central. Locomotive availability and building of Cross Harbour rail link in Belfast may dictate changes. Telephone 09603 53567 for details.

Republic of Ireland-based events, 7 September (Rosslare-Limerick); 8 September (return).

Other day drips often arranged at short notice — contact RPSI, 33 Torquay Wood, Dublin 18 for Republic of Ireland trips and RPSI, Castleview Road, Whitehead BT38 9NA for Northern Ireland trips
Feature article reference: RW October 1989, p594 (1989 Mount Brandon railtour)

Right:
The RPSI's railtour on 23 June 1990 took No 4 through Enfield, Northern Ireland. *A. Eaten*

Shanes Castle Railway SC

A 3ft gauge line runs from the Lodge Gates to the Old Castle ruins through the park and along the shore of Loch Neagh and runs through Lord O'Neill's nature reserve
Headquarters: Antrim station, Shanes Castle Estate, Co Antrim, Northern Ireland
Telephone: Antrim (084 94) 63380/ 62216
Main station: Antrim (SCR)
Car parks: Shanes Castle Estate, Antrim
Access by public transport: By Ulsterbus from several centres. By NI Railways to Antrim station
Refreshment facilities: Shanes Castle station
Length of line: 1½ miles, 3ft gauge
Souvenir shop: Antrim station
Passenger trains: Steam and diesel-hauled trains from Antrim station through the Shanes Castle estate. Diesel service only on Saturdays in May, June and September
Period of public operation: Sundays and Bank Holidays April; Saturdays, Sundays and Bank Holidays in May

Locomotives

Name	*No*	*Builder*	*Type*	*Built*
Tyrone	1	Peckett (1026)	0-4-0T	1904
—	2	Simplex (11039)	0-4-0DM	1956
Shane	3	Barclay (2265)	0-4-0WT	1949
Blue Circle	4	Simplex (102T016)	0-4-0DH	1976
Nancy	5	Avonside (1547)	0-6-0T	1908
Rory	6	Simplex (102TO07)	0-4-0DH	1974

Stock
12 4-wheel carriages
3 4-wheel Tram cars
2 ex-Londonderry & Lough Swilly Railway wagons
1 ex-Belfast & Northern Counties Railway wagon

and September; Wednesday, weekends and Bank Holidays in June, Tuesday, Wednesday, Thursdays and weekends July-August; Sundays in September. Opening times 12.30-18.30
Special events: Contact for details
Facilities for disabled: Limited, no toilets, ramps available
Special notes: Railway serves visitors to the Shanes Castle estate of Lord O'Neill which includes an extensive nature reserve

Ulster Folk & Transport Museum M

This former weaving machinery manufacturer's premises has all the atmosphere of an early Victorian railway terminus with the trains waiting in the platforms. A real Aladdin's cave and alone would justify a visit to the province
Location: Witham Street, Belfast
Operating organisation: Ulster Folk & Transport Museum, Cultra Manor, Holywood, Co Down
Telephone: Belfast (0232) 451519
Access by public transport: Citybus routes 16, 21, 24, 25, 76 and 77 to Rope Works
Car park: No facilities
On site facilities: Shop, toilets. 7¼in line (at Cultra)
Public opening: Daily (except Sunday) 10.00 to 17.00

Locomotives (5ft 3in)

Name	*No*	*Origin*	*Class*	*Type*	*Built*
—	93	GNR(I)	JT	2-4-2T	1895
—	30	BCDR	I	4-4-2T	1901
Dunluce Caste	74	LMS(NCC)	U2	4-4-0	1924
Maeve	800	GSR	B1A	4-6-0	1939
—	1	R/Stephenson (2738)	—	0-6-0T	1891
Merlin	85*	GNR(I)	V	4-4-0	1932

Locomotives (narrow gauge)

Name	*No*	*Origin*	*Class*	*Type*	*Built*
Kathleen	2	CLR	—	4-4-0T	1887
Blanche	2†	CDJRC	—	2-6-4T	1912
Phoenix	11	CVR	—	4wD	1928
—	2	Industrial	—	0-4-0	1956
—	20	Industrial	—	0-4-0	1905

*On loan to RPSI for operational use
†On loan to Derry Railway Heritage Centre

Stock
1 Dublin, Wicklow & Wexford Railway coach
1 Dundalk, Newry & Greenore Railway coach
1 Midland & Great Western Railway director's saloon (ex-private vehicle)
1 Electric tramcar of Bessbrook-Newry Tramway
1 Covered 'toastrack' of Giant's Causeway, Portrush and Bush Valley Railway & Tramway Co
1 Cavan-Leitrim Railway coach
1 County Donegal Railway railcar (plus 1 on loan to DRHC)
1 County Donegal Railway director's coach
1 County Donegal Railway trailer coach (bodywork ex-Dublin & Lucan Railway coach)

Westrail

Location: Tuam, Co Galway, Ireland
OS Reference: Lat 42, Long 52
Operating society/organisation: Westrail (Tuam) Ltd, The Railway Station, Vicar Street, Tuam, Co Galway
Telephone: (093) 24200, (091) 91039, (091) 65269. **Fax:** (091) 65384
Car park: On site
Access by public transport: Irish Rail — Athenry. Bus Eireann — Tuam
Catering facilities: On train, licensed snack bar. Platform shops at Athenry
Length of line: Athenry-Tuam, 15 miles
Period of public operation: Weekends June-August — please check for specific details

Locomotives

Name	*No*	*Origin*	*Class*	*Type*	*Built*
—	90	GSWR	J30	0-6-0T	1875
—	E428	CIE	421	C	1962
—	G613*	CIE	611	B	1962
—	3	CSET	—	0-4-0DM	1960
—	B104	CIE	101	A1A-A1A	1956
—	B113	CIE	113	Bo-Bo	1951

*On loan to Downpatrick & Ardglass Railway

Rolling stock
3 ex-CIE coaches

Journey time: 1-3hr return (see timetable for details)

Memberships details: Contact above address for details

Channel Islands & Isle of Man

Alderney Railway

Location: Alderney, Channel Islands
Operating society/organisation: Alderney Railway Society, PO Box 75, Alderney, Channel Islands
Telephone: 0481-823580
Car park: Yes
Access by public transport: Aurigny Air Services from Southampton and Bournemouth. Torbay Seaways from Torquay
On site facilities: Shop at Braye Road, refreshments (Sundays)
Public opening: Weekends and Bank Holidays, Easter to October
Special events: Alderney Week 3-10 August. Easter Egg Specials on Easter Saturday. Santa Specials, Saturday before Christmas

Industrial locomotives

Name	*No*	*Builder*	*Type*	*Built*
J. T. Daly	3	Bagnall (2450)	0-4-0ST	1931
Elizabeth	D100	Vulcan (D100)	0-4-0DM	1949

Stock

3 Wickham trolleys run as multiple-unit
2 Goods wagons
2 ex-London Underground 1938 Stock tube cars, Nos 10177/11177 (locomotive-hauled)
2 Wickham Flats

Length of line: 2 miles
Facilities for disabled: Yes

Groudle Glen Railway SC

Location: Groudle Glen Railway, Isle of Man
Operating company: Groudle Glen Railway Ltd (managed by the Isle of Man Steam Railway Supporters' Association) of 19 Ballabrooie Grove, Douglas, Isle of Man
Telephone: 0624 622138 (evenings)
Car park: Yes
Access by public transport: Manx Electric Railway
On site facilities: Sales shop
Length of line: ½-mile, 2ft gauge
Public opening: Sundays and Bank Holidays May to September with some Wednesday evening services in July and August; Santa Trains in December
Facilities for disabled: Due to the line's location, those who are disabled will have some difficulty. It is suggested that they telephone for advice
Further information and membership details: From above address

Locomotives

Name	*No*	*Builder*	*Type*	*Built*
Dolphin	1	H/Hunslet (4394)	4wDM	1952
Walrus	2	H/Hunslet (4395)	4wDM	1952
Sea Lion	—	Bagnall (1484)	2-4-0T	1896

Isle of Man Railway

The 3ft gauge Isle of Man Railway is a survivor of a system which covered the whole island. Almost continuous operation since 1874 makes it one of the oldest preserved railways in the British Isles. The railway has changed little since the turn of the century and retains much of its Edwardian atmosphere. It runs for 15 miles between Douglas and Port Erin through the island's rolling southern countryside. One of the original locomotives built in 1874 is still in daily service making it one of the oldest operational steam locomotives in the British Isles

Headquarters: Isle of Man Railways, Strathallan Crescent, Douglas, Isle of Man
Telephone: Douglas (0624) 663366
Main station: Douglas
Other public stations: Port Soderick, Santon, Castletown, Port Erin, Ballasalla, Port St Mary
Car parks: Douglas, Ballasalla, Castletown, Port Erin
Access by public transport: Isle of Man Transport buses to main centres
Refreshment facilities: Port Erin and Douglas
Souvenir shops: Douglas and Port Erin
Museum: Port Erin
Depot: Douglas
Length of line: 15¼ miles, 3ft gauge
Passenger trains: Douglas-Port Erin
Period of public operation: Daily 29 March-5 April and 29 April-28 September

Locomotives

Name	*No*	*Builder*	*Type*	*Built*
Loch	4*	B/Peacock (1416)	2-4-0T	1874
G. H. Wood	10	B/Peacock (4662)	2-4-0T	1905
Maitland	11*	B/Peacock (4663)	2-4-0T	1905
Hutchinson	12*	B/Peacock (5126)	2-4-0T	1908
Kissack	13*	B/Peacock (5382)	2-4-0T	1910
—	19*	Walker (GNR (I))	diesel railcar	1950
—	20*	Walker (GNR (I))	diesel railcar	1951

*Operational, No 10 is stored out of use, not on display.

On display in museum at Port Erin

Name	*No*	*Builder*	*Type*	*Built*
Sutherland	1	B/Peacock (1253)	2-4-0T	1873
Caledonia	15	Dubs & Co (2178)	0-6-0T	1885
Mannin	16	B/Peacock (6296)	2-4-0T	1926

Owned by Isle of Man Railway Society (not on display)

Name	*No*	*Builder*	*Type*	*Built*
Mona	5	B/Peacock (1417)	2-4-0T	1874
*Tynwald**	7	B/Peacock (2038)	2-4-0T	1880
Fenella	8	B/Peacock (3610)	2-4-0T	1894
Douglas	9	B/Peacock (3815)	2-4-0T	1896

*Chassis only

Facilities for disabled: Level access throughout Douglas and Port Erin stations including snack bar. Carriages able to carry wheelchairs, ramps provided. Advance notice helpful

Manx Electric Railway

The 3ft gauge Manx Electric Railway is a unique survivor of Victorian high technology. A mixture of railway and tramway practice, it was built in 1893 and was a pioneer in the use of electric traction. Two of the original cars are still in service making them the oldest tramcars still in operation in the British Isles. After leaving Douglas, the railway passes the newly opened Groudle Glen Railway before reaching the charming village of Laxey, home of the Snaefell Mountain Railway. The line continues over some of the most breathtaking coastal scenery in the

Motor Cars

Nos	*Type*	*Seats*	*Body*	*Built*
1, 2	Unvestibuled saloon	34	Milnes	1893
5, 6, 7, 9	Vestibuled saloon	32	Milnes	1894
14, 15, 17, 18	Cross-bench open	56	Milnes	1898
16	Cross-bench open	56	Milnes	1898
19-22	Winter saloon	48	Milnes	1899
23	Centre-cab locomotive	—	IOMT & EP	1900
25-27	Cross-bench open	56	Milnes	1898
28-31	Cross-bench open	56	ERTCW	1904
32, 33	Cross-bench open	56	UEC	1906

island before reaching its terminus at Ramsey nearly 18 miles from Douglas
Headquarters: Isle of Man Railways, Strathallan Crescent, Douglas, Isle of Man
Telephone: Douglas (0624) 663366
Main station: Douglas (Derby Castle)
Other public stations: Laxey, Ramsey
Car parks: Douglas, Laxey, Ramsey nearby
Access by public transport: Isle of Man Transport buses to main centres
Depots: Douglas, Laxey, Ramsey
Refreshment facilities: Laxey
Museum and souvenirs: Ramsey
Length of line: 17 miles, 3ft gauge
Passenger service: Douglas-Ramsey

Trailers

13	Cross-bench open	44	Milnes	1893
36, 37	Cross-bench open	44	Milnes	1894
40, 41, 44	Cross-bench open	44	EE Co	1930
42, 43	Cross-bench open	44	Milnes	1903
45-48	Cross-bench open	44	Milnes	1899
49-50, 53, 54	Cross-bench open	44	Milnes	1893
52	pw flatcar (ex trailer)	—	Milnes	1893
55, 56	Cross-bench open	44	ERTCW	1904
57, 58	Saloon	32	ERTCW	1904
59	Special Saloon	18	Milnes	1895
60	Cross-bench open	44	Milnes	1896
61, 62	Cross-bench open	44	UEC	1906

Period of public operations: Daily 25 March-6 April; Monday-Saturday 8-27 April; daily 29 April-28 September

Special notes: Folded wheelchairs can be carried. Please notify in advance

Ramsey Electric Railway Museum M

Location: Manx Electric Railway Station, Parsonage Road, Ramsey, Isle of Man
Operating society/organisation: (General enquiries) Isle of Man Railway Society, 4 Clifton Road, Rugby, Warwickshire (Applications for special opening) c/o Isle of Man Railways, Douglas, IoM
Telephone: (General enquiries) 0788 543026
Car parks: Locally in nearby street
Access by public transport: Manx Electric Railway station adjoins museum; Douglas-Laxey-Ramsey buses pass within a few yards, alight at Parsonage Road
On site facilities: Museum shop, toilets in MER station
Public opening: Monday to Friday only; late May to mid-September 11.00-15.45

No	*Origin/Type*	*Built*
	Manx Electric Railway	
23	MER Bo-Bo electric loco	1900
26	MER Bogie freight car	1895
	Douglas Horse Tramway	
11	Douglas, open toastrack	1886
47	Douglas, bulkhead car	1911
49	Douglas, convertible car	1935
—	Horse tram transporter wagon	c1930
	Queens Pier Tramway	
—	Hibberd 'Planet' 4WPM (2037)	1937
—	Hibberd-PRV bogie trailer (2038)	1937
—	4w Skip, ex-Ballajora Quarry	
—	4w Skip, ex-Poortown Quarry	

Special notes: For parties travelling by MER, opening outside normal hours may sometimes be possible

Snaefell Mountain Railway

The 3ft 6in gauge Snaefell Mountain Railway is unique. It is the only electrically-driven mountain railway in the British Isles. Almost all the rolling stock is original and dates back to 1895. The railway begins its journey at the picturesque village of Laxey where its terminus is shared with the Manx Electric Railway. The climb to the summit of Snaefell (2,036ft) is a steep one and the cars travel unassisted up gradients as steep as 1 in 12. From the summit, the views extend to Wales, Scotland, England and Ireland
Headquarters: Isle of Man Railways, Strathallan Crescent, Douglas, Isle of Man
Telephone: Douglas (0624) 663366
Main station: Laxey
Other public stations: Bungalow, Summit
Car parks: Laxey, Bungalow (nearby)

Nos	*Type*	*Seats*	*Body*	*Built*
1-4, 6	Vestibuled saloon	48	Milnes	1895
5 (rebuild)	Vestibuled saloon	48	MER/ Kinnin	1971

Access by public transport: Manx Electric Railway or Isle of Man Transport bus to Laxey
Refreshment facilities: Laxey, Summit
Museum: Ramsey
Souvenirs: Summit
Length of line: 5 miles, 3ft 6in gauge
Passenger service: Laxey-Snaefell summit
Period of public operation: Daily 29 April-28 September
Special notes: No special facilities for disabled

Miniature Railways

Audley End Railway
Audley End, Saffron Walden, Essex. Tel: (0799) 22345 or 27956
Opening details: Daily — Easter week, summer half-term, summer school holidays and Sundays and bank holidays April-October (from 14.00)
10¼in gauge; 1½-miles long; 3 steam, 2 diesel locomotives
Public access: BR Audley End (1 mile), free car park
Site facilities: Light refreshments, toilets
Note: (Postal address) Estate Office, Audley End, Saffron Walden, Essex CB11 4JG

Great Cockcrow Railway
Hardwick Lane, Lyne, Chertsey, Surrey. Tel: Mon-Fri (0932) 228950; Sun (0932) 565474
Opening details: Every Sunday May to October inclusive, 14.15-17.30
7¼in gauge; normal run ⅞-mile; 14 steam locomotives, 1 electric, 1 petrol (nine normally in service)
Public access: BR Chertsey (1¼-miles); Green Line 716 Holloway Hill (½-mile), free car park
Site facilities: Toilet; light refreshments

Lightwater Valley Theme Park
North Stainley, Nr Ripon, North Yorkshire. Tel: (0765) 85368 and 85321 (office)
Opening details: 24hr answer phone on above number
15in gauge; 1-mile long; 4 steam, 1 diesel, 1 petrol locomotives
Public access: BR Harrogate (12 miles), BR Thirsk (9 miles), free car park
Site facilities: Restaurant, food bar, coffee shop, theatre, toilets (including disabled), shops, selection of rollercoaster rides and other attractions

Moors Valley Railway
Horton Road, Ashley Heath, Nr Ringwood, Dorset. Tel: (0425) 471415
Opening details: Open every weekend throughout the year, plus half-term holidays and daily from Spring Bank Holiday to end September. Santa Specials in December
7¼in gauge; 1-mile long; 6 steam locomotives
Public access: BR Christchurch, free car park
Site facilities: Toilets (including disabled), shop

Below:
An undated scene on the Audley End Railway.

Members of the Association of Railway Preservation Societies

Full Members

Bitton Railway Co Ltd: Mr G. Clark, 12 Cranwell Grove, Whitchurch, Bristol BS14 9QR

Bluebell Railway Preservation Society: Mr P. Thomas, 'Grey Friars', Gravelye Lane, Lindfield, Haywards Heath, Sussex

Bowes Railway Co Ltd: Mr W. Craddock, 112 Donvale Road, Donwell, Washington, Tyne & Wear NE37 1DN

The Bulleid Society Ltd: Mr D. A. Foale, 'Namron', South Chailey, Lewes, East Sussex BN8 4AD

Cambrian Railways Society Ltd: Mr J. L. D. Price, 'Delamere', Old Chirk Road, Gobowen, Oswestry, Salop SY11 1PE

Caerphilly Railway Society Ltd: Mr A. Smith, 51 Worcester Crescent, Newport, Gwent

Chasewater Light Railway & Museum Co: Mr B. Bull, Brownhills West Station, Hednesford Road, Brownhills West, Walsall WS8 7LT

Cholsey & Wallingford RPS: Mr D. Button, 1 Warwick Close, Abingdon, Oxford OC10 0NF

Colne Valley Railway Co Ltd: Mr J. R. Hymas, Castle Hedingham Station, Yeldham Road, Castle Hedingham, Halstead, Essex CO9 3DZ

Corris Railway Society: Mr S. Parr, 27 Ivis Close, Selly Oak, Birmingham B29 5BS

Darlington Railway Preservation Society: S&DR Goods Depot, Station Road, Hopetown, Darlington DL3 6ST

Dart Valley Railway Association: Mr M. Webb, 7 The Orchard, Corn Park Road, Abbotskerwell, Devon TQ12 5QE

Dean Forest Railway Society Ltd: Mr C. A. Bladon, Northwood House, Northwood Green, Westbury-on-Severn, Gloucestershire GL14 2NA

Dinting Railway Centre Ltd (Bahamas Locomotive Society): Mr K. J. Tait, 73 Derby Road, Heaton Moor, Stockport, Cheshire SK4 4NG

East Lancashire Railway Preservation Society: Mr L. Vevers, Bury Transport Museum, Castlecroft Road, Bury, Lancashire BL9 0LN

East Somerset Railway Co Ltd: Mr B. Buckfield, The Railway Station, West Cranmore, Shepton Mallet, Somerset

Festiniog Railway Society Ltd;: Mr D. Gordon, 72 West Street, Comberton, Cambridge CB3 7DS

Foxfield Light Railway Society Ltd: Mrs L. Reed, 63 Maythorne Road, Blurton, Stoke-on-Trent ST3 3AE

Friends of the National Railway Museum: Mr J. Spencer Gilks, c/o The National Railway Museum, Leeman Road, York YO2 4XJ

Great Central Railway Co: The Chairman, Loughborough Central Station, PO Box 33, Loughborough, Leicestershire LE11 1RW

Great Western Society Ltd: Mr J. B. O'Hagen, GWS, Didcot Railway Centre, Didcot, Oxon OX11 7NJ

Gwili Railway Co Ltd: The Chairman, 5a Coley Hill, Reading RG1 6AE

H Class Trust: Mr E. R. C. Oades, 5 Senlac Way, St Leonards-on-Sea, Sussex

Irish Steam Preservation Society Ltd: Mr F. C. Flewitt, 6 Waterloo Avenue, Dublin 3, Ireland

Keighley & Worth Valley Preservation Society: Mr R. G. Mitchell, 1 Smithy Row, Hurst Green, Blackburn, Lancs BB6 9QA

Lakeside Railway Society: Mr T. D. Owen, 83 Camberton Road, Linslade, Leighton Buzzard, Bedfordshire LU7 7UW

Leighton Buzzard Narrow Gauge Railway Society Ltd: Miss J. Churchill, 63 Holmbridge Gardens, Enfield, Middx EN3 7EY

Llangollen Railway Society: Mr A. Terry, 33 Deeside, Ellesmere Port, S. Wirral, Cheshire

Locomotive Club of Great Britain: Mr R. L. Patrick, 8 Wolviston Avenue, Bishopgate, York YO1 3DD

Locomotive Owners Group (Scotland) Ltd: Mr J. L. Stevenson, 4 Queens Road, Blackhall, Edinburgh EH4 2BY

Maid Marian Locomotive Fund: Mr H. Jones, 'Walden', 139 Stoops Lane, Bessacar, Doncaster DN4 7RG

Main Line Steam Trust Ltd: Mr S. Jacobs, 'Wayside', Rugby Road, Swinford, Lutterworth, Leics LE17 6BW

Merchant Navy Locomotive Preservation Society Ltd: Mr R. Abercrombie, 12 Inglewood Avenue, Heatherside, Camberley, Surrey

Mid-Hants Railway Preservation Society: Mr R. P. C. Higgs, 18 Barnsford Crescent, West End, Woking, Surrey GU24 9HX

Middleton Railway Trust Ltd: Mr I. Smith, The Station, Moor Road, Leeds LS10 2JQ

Midland & Great Northern Joint Railway Society Ltd: The Librarian, Sheringham Station, Sheringham, Norfolk NR26 8RA

Midland 4F Preservation Society: Mr I. Johnson, 13 Shepley Close, Hazel Grove, Stockport, Cheshire SK7 6JJ

Midland Railway Trust Ltd: Mr R. Clegg, 9 Hawthorn Crescent, Findern, Derby DE6 6AN

Nene Valley International Steam Railway: Wansford Station, Stibbington, Peterborough PE8 6LR

North Eastern Locomotive Preservation Group: Mr D. F. Martin, 21 Grassington Avenue, Northallerton, North Yorkshire DL7 8SY

North Gloucestershire Railway Co Ltd: Mr R. H. Wales, 'Wellesbourne', Oakfield Street, Tivoli, Cheltenham, Gloucestershire GL33 8HR

North York Moors Historical Railway Trust: Mr J. H. Meredith, 11 College Road, Copmanthorpe, York YO2 3US

Quainton Railway Society: Mr R. B. Millar, 'Ducal', 25 Loudham Road, Little Chalfont, Amersham, Buckinghamshire

Railway Preservation Society of Ireland: Mr P. McCann, 19 Ferndale Road, Newtonabbey, Co Antrim BT36 8AH

Ravenglass & Eskdale Railway Preservation Society Ltd: Mr D. Pickup, The Retreat, Ravenglass, Cumbria CA18 1SW

Ruislip Lido Railway Society: Mr R. P. Shafe, 3 Oxleay Road, Rayners Lane, Harrow, Middx HA2 9UZ

Scottish Railway Preservation Society: Dr A. Glen, 21 Monks Road, Airdrie, Lanarkshire ML6 2QW

Severn Valley Railway Co Ltd: Mr M. J. Draper, 'Dyniatts', The Hill, Abberley, Worcester WR6 6BZ

Shackerstone Railway Society Ltd: Mr R. A. Briggs, 40 Newstead Avenue, Burbage, Hinckley

Sittingbourne & Kemsley Light Railway Ltd: Mr M. Burton, 85 Balmoral Road, Gillingham, Kent ME7 4QE

South Tynedale Railway Preservation Society: Mr S. G. Dyke, 11 Norley Avenue, Sunderland SR5 1JY

Southern Electric Group: Mr P. Stavely, 247 Davidson Road, Croydon CR0 6DR

Stanier 8F Locomotive Society Ltd: Mr D. R. McIntosh, 296 Lower Hillmorton Road, Rugby, Warwickshire CV21 4AE

Steam in Hereford Ltd (6000 Locomotive Association): Mr B. R. Wallis, 8 Little Birch Croft, Whitchurch, Bristol BS14 0JB

Stour Valley Railway Preservation Society: Mr L. Houghton, Old Railway Tavern, Station Road, Wakes Colne, Colchester, Essex CO6 2DS

Strathspey Railway Association: Mr R. Black, 37 Castle Avenue, Balloch, Alexandria, Dunbartonshire G83 8HU

Talyllyn Railway Co Ltd: Mr D. Woodhouse, Wharf Station, Tywyn, Gwynedd LL36 9EY

Tenterden Railway Co Ltd (Kent & East Sussex Railway): Mr M. Toynbee, Tenterden Railway Co Ltd, Tenterden Town Station, Tenterden, Kent TN30 6HE

Vintage Carriages Trust: Mr M. W. Cope, Haworth Station, Keighley, West Yorkshire BD22 8NJ

Wainwright 'C' Preservation Society: Mr I. Demaid, 69 Bromley Gardens, Bromley, Kent BR2 0ES

Welshpool & Llanfair Light Railway Preservation Co Ltd: Llanfair Station, Llanfair Caereinion, Powys SY21 0SE

Wight Locomotive Society: The Secretary, Railway Station, Haven Street, Ryde, Isle of Wight PO33 4DS

Yorkshire Dales Railway Museum Trust: Mr M. D. Cleaver, 26 Birch Hall Lane, Earby Colne, Lancashire BB8 6JX

1247 Society: Capt W. G. Smith, 1 Church Road, Willington, Bedford MK44 3QD

1708 Locomotive Preservation Trust Ltd: Mr G. W. Kingham, Registered Office, 106 Stanford Road, Luton, Beds LU2 0QA

6201 Princess Elizabeth Society Ltd: E. J. Whitlock, 4 Lypiatt View, Bussage, Stroud, Gloucestershire LG6 8DA

Associate Members

A4 Locomotive Society Ltd: Mr G. R. Pope, Secretary, Keeper's Cottage, Muntham Farm, North End, Findon, Worthing BN14 0RQ

Avonside Steam Preservation Society: B. Barham, The Buffers, High Street, Hinxton, Saffron Walden, Essex

Ayrshire Railway Preservation Group: Mr G. Thompson, 4 Kyle Crescent, Loans, Troon, Ayrshire KA10 7EZ

Battle of Britain Locomotive Preservation Society: Mr R. J. Tanner, 317 Cardington Road, Bedford MK42 0DU

Bodmin & Wenford Railway plc: Mr R. G. Fitzgerald, Bodmin General Station, Bodmin, Cornwall PL31 1AQ

BRC&W Type 3 Preservation Group: Mr I. Simmons, 131 Sompting Road, Worthing, West Sussex BN14 9EX

Brechin Railway Preservation Society: The Station, 2 Park Road, Brechin, Angus DD9 7AF

Brecon Mountain Railway Association: The Chairman, Pant Station, Merthyr Tydfil, Mid Glamorgan

Bressingham Steam Engine Trust: Mr A. Bloom, Bressingham Hall, Bressingham, Diss, Norfolk IP22 2AB

Britannia Locomotive Society: Mr A. A. Lawson, Gaylands, 9 Bennett Drive, Myton Grange, Warwick CV34 6QJ

Butetown Historic Railway Society Ltd: Mr D. J. Morgan, 34 Bryngwyn Road, Cyncoed, Cardiff CF2 6PG

Cadeby Light Railway: Mrs A. Boston, The Old Rectory, Cadeby, Hinckley, Nuneaton, Warwickshire

Camelot Locomotive Society: Mr P. W. Gibbs, 3 Garden Close, Northolt, Middx UB5 5ND

Conwy Valley Railway Museum Ltd: c/o Burtons of Walsall Ltd, 65 Bradford Street, Walsall, West Midlands WS1 3QD

Cornish Steam Locomotive Preservation Society Ltd: Mr M. Orme, 3 Jubilee Terrace, Goonhavern, Truro, Cornwall TR4 9JY

Cotswold Steam Preservation Ltd: Mr J. M. Colley, Birchwood Farm, Bridgnorth Road, Shatterford, Nr Bewdley, Worcs DX12 1TP

Derwent Railway Society: Mr S. Buck, Owsen Place, Lamplough, Workington CA14 4SO

Diesel and Electric Group: Mr R. E. Tiller, 40 Newbarn Park Road, Taunton, Somerset TA1 4NP

Dyfed Railway Co Ltd: The Chairman, Henllan Station, Llandyssul, Dyfed

Eastleigh Railway Preservation Society: Mr D. B. Smith, 17 Twiggs End Close, Locksheath, Southampton, Hants SO3 6ET

Fairbourne & Barmouth Railway: Mr J. Ellerton, Plas Gwyntog, Rhoslefain, Tonwy, Gwynedd LL37

Forest Pannier Tank Fund: Mr J. S. Metherall, 15 Sudbrook Way, Gloucester

Forest Prairie Fund: Mrs H. Davies, 55 Victoria Street, Cinderford, Glos GL14 2ET

Foxcote Manor Society: Mr I. N. Travers, 5 Harlow Avenue, Upton-by-Chester, Cheshire CH2 1NQ

Gloucestershire Warwickshire Steam Railway PLC: Toddington Station, Toddington, Cheltenham, Gloucestershire GL54 5DT

Gravesend Railway Enthusiasts Society: Mr D. Hanger, 47 Lower Range Road, Gravesend, Kent DA12 2QL

Great Central Railway Coach Group: Mr K. H. Brooker, Manor Road, Sulgrave, Banbury, Oxon OX17 2RY

Great Western Preservations Ltd: Mr A. R. Machon, 15 Calder Close, Maidenhead, Berkshire SL6 7RS

The Gresley Society: Mr G. Goslin, 8 Pevensey Grove, Flitwick, Bedford MK45 1SD

GWR 813 Preservation Fund: Mr P. Goss, 23 Hatchmere, Thornbury, Bristol BS12 3EU

Hastings Diesel Preservation Ltd: Mr N. Carter, 40 Westwood Park, Forest Hill, London SE23 3QH

Hexham Rolling Stock Group: Mr G. Hall, 21 Birkdene, Stocksfield, Northumberland NE43 7EN

Hull & Barnsley Railway Stock Fund: Mr A. Halman, 6 Chequerfield Court, Chequerfield Avenue, Pontefract, West Yorkshire WP8 2TQ

Ivatt Locomotive Trust: Mr R. B. Miller, 'Ducal', 25 Loudham Road, Little Chalfont, Amersham, Buckinghamshire

Lambton No 29 Syndicate: Mr J. M. Richardson, 5 Ravine Hill, Filey, North Yorkshire YO14 9EU

Lancashire & Yorkshire Railway Preservation Society: Mr A. Cox, 3 Bastwell House, Blackburn, Lancashire BB1 9TY

Lincolnshire Coast Light Railway: Mr H. L. Goy, 12 Giles Street, Cleethorpes DN35 8AE

Liverpool Locomotive Preservation Group: Mr K. Soper, 90 Brick Kiln Lane, Rufford, Lancs L40 1SY

Lloyds Railway Society: Hon Secretary, Lloyds, Lime Street, London EC3

LM 2MT 46464 Trust: Mr I. N. Fraser, Palace Gates, Viewfield Road, Arbroath, Angus DD11 2BS

London & North Western Society: Mr J. C. James, 'Solaby', 4 Longview Drive, Huyton, Liverpool L36 6EE

Lowthers Railway Society Ltd: Mr D. Lappin, 44 Lednock Road, Stepps, Glasgow G33 6LU

Manchester Rail Travel Society: Mr F. G. Cronin, 12 Chiltern Road, Ramsbottom, Lancashire

Manston Locomotive Preservation Society: T. L. Mann, 41 Crow Hill Road, Garlinge, Margate, Kent CT19 5PF

Maunsell Locomotive Society: Mr R. Packham, Chairman, 132 Church Road, Swanscombe, Kent DA10 0PH

North Staffordshire Railway Co Ltd: Mr A. Dale, The Railway Station, Cheedleton, Staffs ST13 7EE

North Tyneside Steam Railway Association: Mr R. J. Maughan, Stephenson Railway Museum Project, Middle Engine Lane, W. Chirton, N. Shields, Tyne & Wear

Peak Rail PLC: Mr R. Raynor, 2A Avenue Road, Forest Gate, London E7 0LD

Plym Valley Railway Association: Mr B. Mills, 11 Hill View, Buckland, Monachorum, Yelverton, Devon PL20 7ND

Pontypool & Blaenavon Railway Society: Mr P. Powell, Haul-y-Bryn, Church Road, Abersychan, Gwent NP4 7EF

Railway Club of Wales: Mr N. Wessell, Aberlogin Cottage, Llanmorlais, Swansea SA4 3TY

Railway Vehicle Preservation Ltd: Mr G. E. Maslin, 25 Peterborough Road, Farcet, Peterborough PE7 3BH

Redditch Steam Locomotive Preservation Society: Mr A. Marsden, 72 Longhurst Croft, West Heath, Birmingham B31 4SQ

Rheilffordd Llyn Tegid Ltd (Bala Lake Railway): Mr G. H. Barnes, 'Yr Orsaf', Llanuwchllyn, Bala, Gwynedd

Romney, Hythe & Dymchurch Railway Association: Mr A. Goyns, 'Fryers', Church Road, New Romney TN8 8EY

Rutland Railway Museum: Mr G. E. Kobish, 9 The Knoll, Great Gonerby, Grantham, Lincolnshire NG31 8JY

Salisbury Steam Locomotive Preservation Trust: Mr E. J. Roper, 33 Victoria Road, Wilton, Salisbury, Wiltshire SP2 0OZ

Somerset & Dorset Railway Trust: M. J. Palmer, The Haven, Chandlers Lane, Edington, Bridgwater, Somerset TA7 9JY

Southern Steam Trust: Mr M. L. Norris, The Station, Swanage, Dorset SA1 1SN

South Yorkshire Railway Preservation Society: Mr J. Wade, 138 Newman Road, Wincobank, Sheffield S9 1LS

Stanier Black Five Locomotive Preservation Society: Mr D. J. Porter, 29 High Street, Shoreham Village, Kent TN14 7TD

Steamport Southport: Mr H. J. M. Royden, 35 Argameols Road, Freshfield, Crosby, Merseyside L37 7BY

Steam Power Trust '65: Mr K. H. Cockerill, 30 Hartburn Lane, Stockton-on-Tees TS18 3QH

Swindon & Cricklade Railway Society: Mr J. Larkin, Boscreeg, Church Lane, Fernham, Faringdon, Oxon

Telford Horsehay Steam Trust: Mr M. Stone, 11 Cuckoo Oak Green, Bridgnorth Road, Madeley, Telford, Shropshire

Thompson B1 Locomotive Trust: Mr J. G. Gwnett, 11 Valance End, Dunstable LU6 3LP

Tunbridge Wells & Eridge RPS: Mr F. Wallace, The Old Dairy, Snatts Lane, Uckfield, East Sussex TN22 2AL

Underground Railway Rolling Stock Trust: Mr D. C. Alexander, 13 Irvine Drive, Stoke Mandeville HP22 5UN

Urie Locomotive Society: Mr A. Ball, 'Lavenham', Adams Lane, Selbourne, Alton, Hants GU34 3LJ

Wells & Walsingham Light Railway: Mr R. W. Francis, Wells-next-the-Sea, Norfolk

Welsh Highland Light Railway 1964 Ltd: Mr W. J. Jones, Avalon, Llanfrothen, Gwynedd LL48 6LJ

Welsh Industrial and Maritime Museum: Dr E. S. Owen-Jones, Bute Street, Docks, Cardiff CF1 6AN

West Somerset Railway Co: The Railway Station, Minehead, Somerset TR24 5BG

Western Locomotive Association: Mr A. M. Clarke, 6 Greatly, Welwyn Garden City, Herts AL7 4TS

Winchcombe Railway Museum Association: Mr T. R. Petchey, 23 Gloucester Street, Winchcombe, Gloucestershire GL54 5LX

Worcester Locomotive Society Ltd: Mr R. N. Pritchard, 31 Highfield Road, Malvern, Worcs WR14 1HR

1857 Society: Mr K. R. Bowen, 18 Lochmore Close, Hollycroft, Hinckley, Leicestershire

4247 Preservation Society: Mr N. Powles, Station House, Lower Heyford, Oxon OX5 3PD

6024 Preservation Society Ltd: Mr E. F. W. Ives, 63 Georges Hill, Widmer End, High Wycombe, Buckinghamshire HP15 6BH

9462 Preservation Group: Mr W. L. Jones, 74 Coychurch Road, Bridgend, Glamorgan CF31 2AP

45428 Stanier Class Five Locomotive Society Ltd: Mr J. B. Hollingsworth, 'Creua', Llanfrothen, Penrhyndeudraeth, Gwynedd LL48 6SH

71000 Duke of Gloucester Steam Locomotive Trust Ltd: Mr J. Shuttleworth, 25 Patiene Road, Clapham, London SW11 2PY

Affiliate Members

ABAC: Dr J. Brenot, Boite No 1, 33034 Cedex Bordeaux, France

ABPF: The Chairman, Caixa Postal 6501 01051, São Paulo, Brazil

AECFM: M J. David, President, Marcilly-sur-Maulne, 37330 Chateau-la-Valliee, France

APPEVA: M Jacques Pradaro, 21 Rue Gueneguard, 75006 Paris 6, France

Berliner Eisenbahn Freunde: Herr Rudiger Reich, Stresemanstrasse 30, 1000, Berlin, Germany

CFM Blonay-Chamby: Mr J. F. Andrist, PO Box 366 CH 1001, Lausanne, Switzerland

Glasgow Museum of Transport: Mr J. Clayson, c/o Art Gallery & Museum, Kelvingrove, Glasgow G3 8AG

Great Western Railway Museum: Mr T. F. Bryan, Faringdon Road, Swindon, Wilts SN1 5BG

Guild of Railway Artists: Mr F. P. Hodges, 45 Dickens Road, Warwick CV34 5NS

Ian Allan Group: Chairman, Mr Ian Allan, Clock House, Station Approach, Shepperton TW17 8AS

Jarrold Colour Publications: The Chairman, Barrack Street, Norwich, Norfolk NR3 1TR

Leicester University Library: Periodicals Department, University Road, Leicester LE1 7RH

London Transport Museum Library: Mrs M. A. Fletcher, 39 Wellington Street, Covent Garden, London WC2E 7BB

Monkswearmouth Station Museum: The Curator, North Bridge Street, Sunderland SR5 1AP

Museum Buurt Spoorweg: M. Livius Kooy, Vincentstraat 15 7481, LB-Haaksbergn, Netherlands

Narrow Gauge Railways of Wales: Mr D. Woodhouse, Wharf Station, Tywny, Gwynedd LL36 9EY

National Federation of Rail Societies: Mr K. Henderson, PO Box 2429, Auckland, New Zealand

National Museum of Science & Technology: The Librarian, PO Box 9724, Terminal T, Ottawa, Ontario K1G 5A3, Canada

National Railway Historic Society: Mr V. A. Vaughan, 320 Wisconsin Avenue, Oak Park, Illinois 60302, USA

National Tramway Museum Society: The Chairman, Crich, Matlock, Derbyshire DE4 5DP

National Trust Industrial Railway Museum: Mr I. W. Jones, Penrhyn Castle, Bangor, Gwynedd LL57 4AH

Pacific Vapeur Club: M Henri Williaume, Boite Postale 115 76303, Sotteveille-les-Rouen, France

Railway Associations in London: Mr J. Heinemann, 43 Barn Rise, Wembley Park, Harrow, Middlesex HA9 9NT

Smalsparet Vajo-H-Vastervik: Philip Groves, WHVJ-Sec, Box 64 14600, Tullinge, Sweden

Stichting Brabant Rail: Mr J. E. F. L. Simons, Karel Dormanlaan 14, 4B19 BK Breda, Netherlands

Stoomscentrum Maldegem: Rik Degruyter, De Streep 19, B-B340 Damme-Sysele, Belgium

Stoompoorlijn Dendermonde-Puurs: Mr L. Blommaerts, Ten Berg 101, B 9300 AALS, Belgium

Stoomtram Goes Borsele: Mr J. P. Wijtenburg, Postbus 250 4460, AR Goes, Netherlands

Stoomtram Hoon-Medemblik: Mr Peter Scholten, PO Box 137, 1620-AC Hoorn, Netherlands

Transport Ticket Society: Mr A. Fairholm, 16 Blake Road, Seafar, Cumbernauld, Glasgow

Transport Trust: Mr D. Muirhead, Room 131, Melbury Houseice, Marylebone, London NW1 6JU

Ulster Folk & Transport Museum: Mr J. S. Moore, Cultra Manor, Hollywood, Co Down, Northern ireland BT18 0EU

Westinghouse Signals: Mr H. R. Preater, PO Box 79, Pew Hill, Chippenham, Wiltshire SN15 1ND

Late Information

Main Line Steam

Locomotives based at Crewe Heritage Centre for use on 'North Wales Coast Expresses' during 1991 are expected to include: Nos 46229 *Duchess of Hamilton* (NRM), 45596 *Bahamas* (K&WVR), 777 *Sir Lamiel* (NRM, on loan), 4472 *Flying Scotsman* (Steamtown), 4771 *Green Arrow* (NRM), 4498 *Sir Nigel Gresley* (Steamtown, when available).

Locomotives based at Steamtown, Carnforth, to work the 'Cumbrian Coast' and Cumbrian Mountain Expresses' during 1991 are expected to include: 48151 and 46203 *Princess Margaret Rose* (Midland Railway Centre), 8233 (Severn Valley Railway), 6201 *Princess Elizabeth* (Bulmer), 60532 *Blue Peter*, (North Yorkshire Moors, when available), and 71000 *Duke of Gloucester* (Didcot).

Details of individual runs, dates etc will be found in *Railway World*.

Locomotives on loan

It is anticipated that the following locomotives will be on loan for part of the 1991 season:

Llangollen: 46643 (from SVR)
North Norfolk: 47383 (from SVR)
North Yorkshire Moors: 34072 (from Swanage)
West Somerset: 6106 (from Didcot)
Kent & East Sussex: 4566 (from SVR)

New additions to stock:

Great Central: 08114 (ex-Swindon & Cricklade)
Nene Valley: German-built 2-10-0 (Kriegslok) No 7173 (ex-Poland)
Northampton Steam: 27024 (ex British Rail)
Steamtown: 5000 (ex-SVR, NRM-owned)

Other notes:

Swanage: Yugoslav 'USA' is No 62-669, to be restored as SR No 30075. 80104 will be restored at Avon Valley Railway and 30053 at East Anglian Railway Museum (both ex Swindon Heritage Centre)
Llangollen: 5199/6430 to undergo restoration at a private site in the Midlands
Ffestiniog: members have acquired *Monarch* (W&LLR) but is expected to remain at Llanfair Caereinion for 1991

Expected transfers

D7018 (Didcot) no location advised
196 *Errol Lonsdale* (Mid-Hants) to Isle of Wight

Association of Railway Preservation Societies Ltd

Company Limited by Guarantee and not having a share capital.

Registered in England No 1222717

(Registered Office: 21 Market Place, Wednesbury, West Midlands WS10 7AY)

Administrative Office: 3 Orchard Close, Watford, Hertfordshire WD1 3DU. Tel: (0923) 221280. Fax: (0923) 241023 (ARPS).

President: Dame Margaret Weston DBE

Vice President: Capt Peter F. Manisty MBE, Royal Navy (Retd) (also Member of Council), 2 Cleaver Square, London SE11 5DW. Tel: 071-735 5012.

Directors/Council of Management

Chairman and Legal Advisor:
David Morgan, 7 Cheyne Place, London SW3 4HH. Tel: 01-352 6077

Vice Chairman and Meetings Officer:
Peter Ovenstone, 33 Palmerston Place, Edinburgh EH12 5AU. Tel: 031-225 1486 (answerphone).

Company Secretary/Treasurer:
Robert Yates FCA, 21 Market Place, Wednesbury, West Midlands WS10 7AY. Tel: (Office) 021-556 1084.

General Administrator:

Technical Officer:
David Madden, The Station, Sheringham, Norfolk NR26 8RA. Tel: (0263) 822045.

Press Officer:
John Jeffery, 42 North Street, Oundle, Peterborough PE8 4AL. Tel: (Office) (0733) 67474, ext 2129, (Home) (0832) 73010.

Annual Award:
John Ransom, Woodside, Craggan, Lochearnhead, Perthshire FK19 8QD.

Coaching Stock:
Chris Smyth, 7 Woodside, Knutsford, Cheshire WA16 8BX. Tel: 0565 51241.

Minutes Secretary and Overseas Liaison:
Richard Tapper, 39 Grange Court, Boundary Road, Newbury, Berkshire RG14 7PH. Tel: (0635) 30464.

Membership Secretary (Corporate):
David Woodhouse, c/o Talyllyn Railway, Wharf Station, Tywyn, Gwynedd LL36 9EY. Tel: (0654) 710472.

Membership Secretary (Private and Membership Records):
Arthur Harding, 6 Ullswater Grove, Alresford, Hants SO24 9RP. Tel: (0962) 733327.

Officers and Advisors

Sales Officer:
Edward Boxell, 17 Azalea Close, Uxbridge Road, Hanwell, London W7 3QA. Tel: (home) 01-840 2917.

Information Papers Officer:
Revd Eric Buck, 40 Chapel Lane, Wymondham, Norfolk NR18 0DL. Tel: (0953) 602153.

Journal Editor:
Rodney Pitt, 1 Queen's Drive, Newport, Shropshire TF10 7EU. Tel/Fax: 0952 811444.

Railway Heritage Awards:
Gordon Biddle, The Crossings, Levens, Kendal, Cumbria. Tel: (05395) 60993.

Historical Adviser:
Dick Riley, 115 Albemarle Road, Beckenham, Kent BR3 2HS.

Diesel/Electrical Adviser:
John Crane, 7 Robert Close, Potters Bar, Herts. Tel: (0707) 43568.

Museum Liaison:
Stephen Dyke, Darlington Railway Museum, North Road Station, Darlington DL3 6ST. Tel: 03255 460532 (office), 091-548 2163 (home).

Publications Adviser:
Roger Cromblehome, 140 New Road, Netherton, Dudley, West Midlands DY2 6AY.

Railway Operating Adviser:
Allan Garraway MBE, Coedwig, Nethybridge Road, Boat of Garten, Invernesshire PH24 3BQ. Tel: 047-983 303.

Railtours Adviser:
Malcolm Burton, 85 Balmoral Road, Gillingham, Kent.

Railway Castings:
Adrian Crafer, 36 Parklands Road, Swindon, Wiltshire SN3 1EG.

Railway Catering:
David Williams, The Station, Sheringham, Norfolk NR26 8RA. Tel: 0263 822045.

Rating:
David Smith, 5 Svenskaby, Orton Wistoe, Peterborough PE2 0ZY. Tel: 0733 239866.

Society Archivist:
Susan Youell, 5 North Grange Mount, Leeds LS6 2BY. Tel: 0532 786441.

Special Projects Adviser:
Andrew Roberts, 2 Litchfield Close, Frog Hall, Brixworth, Northampton NN6 9BP.

Technical Adviser:
John Snell, RH&DR Head Office, New Romney, Kent TN28 8PL.

Magazine Competition Organiser:
David Wilson, 19 Manor Close, Rothwell, Leeds, West Yorkshire LS26 0RF. Tel: 0532 824147.

Chairman Irish Branch:
John Lockett, 21 Knollwood Road, Branbridge, County Down BT32 4PE.

To a person actively interested in nationwide railway preservation as opposed to one particular preservation scheme, PRIVATE MEMBERSHIP of the ASSOCIATION OF RAILWAY PRESERVATION SOCIETIES LTD offer many advantages. Three major meetings are organised annually, often at the railway site of a leading Member Society. Here one can meet well known personalities in the railway preservation world, and the host society invariably lays on a full day's programme which is both stimulating and enjoyable.

The ARPS Journal is sent to members, containing mostly the latest news of member societies' activities but also full of interesting titbits such as lists of steam locomotives for sale.

Railways Restored and other publications by Ian Allan in conjunction with ARPS are made available at reduced prices.

Private Membership Application Form (1991)

Send this form (or photostat of) to: Membership Secretary, ARPS, 6 Ullswater Grove, Alresford, Hants SO24 9NP

Name..
(BLOCK LETTERS PLEASE)

Address...

....

Post Code.. Telephone.

Subscription enclosed .... Donation enclosed.

Private membership costs: 1 April-31 March (UK addresses) £10, (overseas) £12; Life membership £120 (UK), £140 (overseas).

Index